ALGEBRA

Sid Thatte

Sid Thatte has an M.B.A. in general management from the University of South Carolina and a B.E. in mechanical engineering from the University of Pune. His professional career has spanned nuclear projects, insurance industry and academic instruction. In his spare time, he volunteers at institutions that benefit everyone. In fair weather, he enjoys hiking in the great outdoors.

ALGEBRA

ISBN-10:   1-5399-6946-0
ISBN-13: 978-1-5399-6946-4

Published in the U.S.A. with CreateSpace Independent Publishing Platform

CONTENTS

PREFACE

As is the case with any type of knowledge, if you don't use it, you lose it. Algebra is no exception to that rule. It's possible you may have studied the subject back in school but have now forgotten it. Or maybe it's your first time studying it. In either case, you made the right choice by trying out this book because I have written it to teach algebra from scratch.

This work is my sincere attempt to formulate a *first book on algebra* for those who feel intimidated by the subject. The only assumption I have made is that you know how to add, subtract, multiply and divide whole numbers. Teaching that is beyond the scope of this book. I have, however, provided a multiplication table at the very end of the book (Appendix B) for quick reference. Please also note that I have not written this book to address advanced algebra.

A study plan I'd recommend is to follow the chapters in the order in which they are written. However, if you decide to skip that sequence, I'd at least ask that you read the *sections* within a chapter in the order in which they show up. The reason is that concepts build on top of one another as you progress, and explanations to problems get more and more to-the-point and streamlined along the way. I have done so to avoid redundancy in explanations. Also, there may be more than one ways to solve an example. To keep the reader from being confused, I have consistently focused on one method for each type of example throughout the book.

While working through this guide, you might notice that I have frequently switched variables from problem to problem instead of always using the variable x. I've done so to eliminate your fear of variables. Variables are simply letters. Letters make us literate. Literacy helps us absorb knowledge: the knowledge of algebra.

Upon solving problems at the end of each section within a chapter, if you need more practice to build your confidence, then I'd recommend searching the internet using phrases such as "quadratic equations practice" or "linear functions practice" or "rational expressions practice" etc. pertaining to the topics you're looking for.

I have written the manuscript, done the typesetting, performed the proofreading and designed the cover of this book. *Phew!* It is my hope that my work meets your expectations. I'd encourage you to write a review of this book on Amazon.com so I can stay informed of my students' experiences.

Thank you for giving me a chance, and good luck with the study!

Sid Thatte, M.B.A.

CHAPTER 1
ARITHMETIC (PRE-ALGEBRA)

So what's a chapter on *arithmetic* doing in a book on *algebra*? That's a really good question, reader. Algebra uses letters from the alphabet, whereas arithmetic uses numbers. But the rules of operation in algebra completely mimic those in arithmetic. Therefore, the need to first study how arithmetic works.

TYPES OF NUMBERS

The very first type of numbers are **natural numbers** used to count things we see in our *natural* surroundings: trees, animals, people, hills etc. A natural number can't be zero or negative, but can only be a positive number with no fraction. Thus, natural numbers are 1, 2, 3, 4, … The numbers 3½, −15, 8¾ etc. aren't natural numbers.

The next category is that of **whole numbers**, which includes all natural numbers plus the number *zero*. Thus, whole numbers are 0, 1, 2, 3, …

Integers include all natural numbers, their negatives and zero. Thus, integers are …, −4, −3, −2, −1, 0, 1, 2, 3, 4, …

A **fraction** is a number expressed as part of a whole. For example, $\frac{2}{5}$ is a fraction used to represent 2 out of 5, or 2 on a scale of 5. The upper number 2 is called **numerator**, and the lower number 5 is called **denominator**. When the numerator is less than the denominator, like in $\frac{2}{5}$, it's called a **proper fraction**. When the numerator is greater than the denominator, like in $\frac{7}{4}$, it's called an **improper fraction**. A third type of fraction is called a **mixed fraction** which has a whole number component and a proper fraction component, for example $1\frac{5}{6}$, meaning $1+\frac{5}{6}$.

Just like a fraction, a **decimal** comprises an integer component and a fractional component. For example, 2.58 means 2 + 0.58. A realistic case of that number may be $2.58, where it's 2 whole dollars (the integer component) and $\frac{58}{100}$ of an additional dollar (the fractional component).

Numbers may also be expressed as **percents** (out of a hundred, for example 65%), **radicals** (square roots, for example $\sqrt{15}$) and **exponents** (squares, cubes etc., for example $(-1.4)^3$).

All the types of numbers discussed thus far are collectively categorized as **real numbers**, and have a place on what's called the **real number line**, or simply the **number line** as shown below.

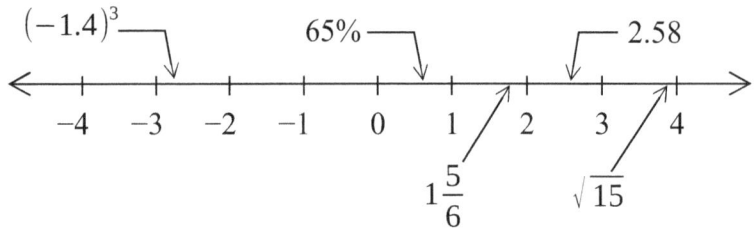

So if there are real numbers, then are there unreal or imaginary numbers as well? Actually, yes. For example, square roots of negative numbers don't exist and are categorized as **imaginary numbers**. Such numbers can't be shown on the number line. We'll study them in Chapter 7 of this book.

SIGNED NUMBERS

As we can see, numbers may have positive or negative signs depending upon which side of zero they belong to on the number line. Let's list the rules for multiplication and division of signed numbers, and then move on to discussing addition, subtraction, multiplication and division of each type of numbers we mentioned up to this point.

MULTIPLICATION OF SIGNED NUMBERS

Positive Number × Positive Number = Positive Number	Symbolically, $(+) \times (+) = (+)$
Positive Number × Negative Number = Negative Number	$(+) \times (-) = (-)$
Negative Number × Positive Number = Negative Number	$(-) \times (+) = (-)$
Negative Number × Negative Number = Positive Number	$(-) \times (-) = (+)$

DIVISION OF SIGNED NUMBERS

Positive Number ÷ Positive Number = Positive Number	$(+) \div (+) = (+)$
Positive Number ÷ Negative Number = Negative Number	$(+) \div (-) = (-)$
Negative Number ÷ Positive Number = Negative Number	$(-) \div (+) = (-)$
Negative Number ÷ Negative Number = Positive Number	$(-) \div (-) = (+)$

You may notice that the rules for multiplication and division of signed numbers are identical.

INTEGERS

The first type of numbers we shall study are **integers**. Recall that integers include all natural numbers, their negatives and the number zero. These numbers don't have fractional components and are therefore easy to deal with. Let's discuss them starting now.

ADDITION/SUBTRACTION OF INTEGERS

When an operation involves only addition or subtraction of integers, perform the tasks from left to right in the order in which they show up. This can be better understood with the following examples.

1.001
10+4+3
14+3   10 and 4 were added first
17   14 and 3 were added next

1.002
−2+5−1
3−1   −2 and 5 were added
2   1 was subtracted from 3

1.003 (−6)+(−8)

As compared with example 1.002, notice the negative numbers written in parenthesis. That's just another way of writing them. But what do we do with the + and the − signs adjacent to each other? Well, **when two signs occur adjacent to each other, consider them as being <u>multiplied</u> with each other**, for which use the rule from the previous page: (+) × (−) = (−)

Therefore, the operation now becomes:
−6−8
−14

When adding or subtracting a number from a <u>negative number</u>, here's a thought that might help make calculations easier: In the above example, look at −6−8. What if you had a bank balance of *negative* $6, and what if you *withdrew* an additional $8 from your bank account? (Let's assume your bank would be nice enough to let you do that.) Well, you'd be making things worse for yourself, because now you'd have an even more negative balance, −$14, which is the equivalent of the two numbers added up together and then written in the negative. See the point?

1.004 −4−(−20)+(−7)−(+3)

First, get rid of the excess signs adjacent to each other by using the rules for multiplication of signed numbers, just like in the previous example.
−4+20−7−3
Now perform operations from left to right, one at a time. Consider a bank balance of negative $4 to which you make a *deposit* (not withdrawal) of $20. Now you're $16 in the positive.
16−7−3
Next, you withdraw $7, and then $3. You'd finally have a bank balance of positive $6.
9−3
6

1.005
2−(−7)+(−1)
2+7−1
9−1
8

1.006
(−14)−(+3)+(+2)−(−5)
−14−3+2+5
−17+2+5
−15+5
−10

PRACTICE PROBLEMS

1.007 21−(+3)

1.008 18+(−7)−(+2)

1.009 −1−(−6)+(−14)

1.010 4−(+4)−(−10)

1.011 −11−(−17)−9

1.012 31+(−26)−(−3)

1.013 −16−(−8)+21

1.014 −6+(−16)−(−26)

MULTIPLICATION OF INTEGERS

When multiplying two or more signed numbers, first decide what the *sign* of the answer will be: whether positive or negative. If all the numbers being multiplied are already positive, then the answer will be positive. **If an <u>odd</u> number of numbers are negative, then their product will be negative**. For example, $(-) \times (-) \times (-) = (-)$. Alternatively, **if an <u>even</u> number of numbers are negative, then their product will be positive**. For example, $(-) \times (-) \times (-) \times (-) = (+)$

Once you've decided the sign, write it down (if negative) and then start multiplying the numbers from left to right, one at a time. The following examples demonstrate the idea.

1.015 $\quad 2 \times 15$
$\qquad 30 \qquad$ Both numbers were already positive, therefore the end result is *positive*.

1.016 $\quad (-4) \times (-8) \times (-2)$
$\qquad$ *Odd number* of negative numbers are being multiplied. The end product will be *negative*. First, write the negative sign. Then simply multiply the numbers from left to right, one at a time.
$\qquad -4 \times 8 \times 2$
$\qquad -32 \times 2$
$\qquad -64$

1.017 $\quad (-9) \times 8$
$\qquad$ Negative times positive gives negative as the end result.
$\qquad -72$

1.018 $\quad (-2) \times (-2) \times (-2) \times (-2)$
$\qquad$ *Even number* of negative numbers are being multiplied. The end result will be *positive*.
$\qquad 4 \times 2 \times 2$
$\qquad 8 \times 2$
$\qquad 16$

PRACTICE PROBLEMS

1.019 $\quad (-3) \times (-9) \times (-1)$ $\qquad$ **1.021** $\quad (-1) \times (-5) \times 2 \times (-1)$ $\qquad$ **1.023** $\quad 4 \times (-3) \times 2$

1.020 $\quad 6 \times (-5) \times (-2)$ $\qquad$ **1.022** $\quad (-8) \times (-3) \times 2$ $\qquad$ **1.024** $\quad (-7) \times 6$

DIVISION OF INTEGERS

When dividing one number by another, first decide on the *sign* of the end result (whether positive or negative) using the rules from page 2. The rules are basically the same for multiplication and division. Then proceed with the actual division.

1.025 $\quad \dfrac{-100}{25}$
$\qquad$ A negative number is being divided by a positive number. The end result will be *negative*.
$\qquad -4$

1.026 $24 \div 8$

Both numbers are already positive. The end result will be *positive*.

3

1.027 $\dfrac{12}{(-2)}$

A positive number is being divided by a negative number. The end result will be *negative*.

−6

1.028 $(-15) \div (-5)$

Both the numbers are negative. The end result will be *positive*.

3

PRACTICE PROBLEMS

1.029 $\dfrac{-60}{12}$

1.030 $(-35) \div 7$

1.031 $(-42) \div (-6)$

1.032 $\dfrac{14}{(-2)}$

1.033 $\dfrac{(-80)}{(-4)}$

1.034 $28 \div (-4)$

EXPONENTS

When a number (positive or negative) is multiplied by itself a certain number of times, that *certain number of times* is called the **exponent**, whereas the original number is called the **base**. For example, $3 \times 3 \times 3 \times 3 = 3^4$ because 3 is multiplied by itself 4 number of times. Here, the 4 is the exponent and the 3 is the base. We read 3^4 as "three raised to four" or "three to the fourth power." As a side note, exponent is also called **power**, **degree** or **index**.

As you can see, expressing numbers in terms of exponents primarily involves multiplication. We'll learn more about exponents in Chapter 2, but for now let's solve a few examples to bolster our understanding of the concept.

1.035 5^2

5 raised to 2 means 5 is multiplied by itself 2 number of times.

5×5

25

1.036 $(-2)^3$

(-2) is multiplied by itself 3 number of times.

$(-2) \times (-2) \times (-2)$

The end result will be *negative*, remember?

$-2 \times 2 \times 2$

−8

1.037 6^3

$6 \times 6 \times 6$

216

1.038 -2^4

Notice that this is not the same as $(-2)^4$. Meaning, the question is asking you to take the *negative of* "two raised to four," not "*negative two* raised to four." So, first multiply 2 by itself 4 number of times. Whatever answer you get, write it as a *negative*.

$-2 \times 2 \times 2 \times 2$

-16

As a side note, $(-2)^4 = (-2) \times (-2) \times (-2) \times (-2) = 2 \times 2 \times 2 \times 2 = 16$

It's the parentheses that make all the difference.

1.039 $(-1)^{200}$

Let's not kid ourselves by expecting to write (-1) times itself 200 number of times! Instead, let's first decide on the sign of the number. It's going to be *positive* because the exponent 200 is an *even* number. And 1 times itself any number of times would still be 1. The final answer is 1.

PRACTICE PROBLEMS

1.040	$(-4)^3$	**1.042**	$(-2)^6$	**1.044**	-1^{50}
1.041	-2^6	**1.043**	$(-5)^3$	**1.045**	$(-1)^{807}$

ORDER OF OPERATIONS

When a problem involves several operations – addition, subtraction, multiplication, division – and maybe some parentheses and exponents thrown in as well, in what order should we prioritize those operations? That's a good question, reader. For that, remember **PEMDAS** – **P**arentheses, **E**xponents, **M**ultiplication, **D**ivision, **A**ddition, **S**ubtraction – in that order of precedence.

Please note that there are **two exceptions** to the PEMDAS rule:
1) When you encounter a patch involving *only addition and subtraction*, do the calculations from left to right in the sequence in which they show up (even if subtraction is *before* addition). Review the start of page 3 if need be.
2) When you encounter a patch with *only multiplication and division*, do the calculations from left to right in the sequence in which they show up (even if division is *before* multiplication).

1.046 $2 + 30 \div (-5)$

In this example and the ones that follow, the ⌢ wave above an operation represents the operation being prioritized in that step. The wave is not a mathematical symbol; it's just my way of drawing your attention to that particular part of the problem.

$2 + \overbrace{30 \div (-5)}$ Division occurs before addition in PEMDAS.

$2 + (-6)$ Now change the two adjacent signs into one: $(+) \times (-) = (-)$

$2 - 6$

-4

1.047 $(-7) \times 4 \div 14 + 9$

$\overbrace{(-7 \times 4)} \div 14 + 9$ In the patch involving only $\times$ and $\div$ go from left to right, doing $\times$ first.

$\overbrace{-28 \div 14} + 9$ Now prioritize $\div$ over $+$ according to PEMDAS.

$$-2+9$$
$$7$$

1.048 $50+(-4)\times8+6$

$\overbrace{50+(-4)}\times8+6$ First, change the adjacent signs into one: $(+)\times(-)=(-)$

$50-\overbrace{4\times8}+6$ Prioritize $\times$ over others according to PEMDAS.

$\overbrace{50-32}+6$ This involves only addition and subtraction. Go from left to right.

$18+6$

24

1.049 $18+\dfrac{(5-3)^2}{2}$

$18+\dfrac{\overbrace{(5-3)}^2}{2}$ First do what's inside the parentheses according to PEMDAS.

$18+\dfrac{\overbrace{2^2}}{2}$ Now do the exponent.

$18+\dfrac{\overbrace{4}}{2}$ Notice that $\dfrac{4}{2}$ is the same as $4\div2$ which means prioritize it.

$18+2$

20

1.050 $5\times((-2)^3+2^4)+10$

$5\times(\overbrace{(-2)^3+2^4})+10$ First calculate what's inside the outer parentheses. Notice that there's no way the exponents can be done *after* parentheses.

$5\times(\overbrace{-8+16})+10$ Simplify further what's inside the parentheses.

$\overbrace{5\times8}+10$ Prioritize $\times$ over $+$

$40+10$

50

PRACTICE PROBLEMS

1.051 $(-7+5)^3\times(-7)\div2$

1.052 $-48\div12\times7+3$

1.053 $(100-49)\div3+\dfrac{-8}{2}$

1.054 $\dfrac{(6+(-3))^3}{9}$

1.055 $4-(-7)\times(-6)+12$

1.056 $6\times(4-5)^{207}+6$

GREATEST COMMON FACTOR (GCF)

The meaning of a **factor** (or divisor) is a number which can *divide* another number. For example, 5 can divide 30. In other words, the number 5 goes into 30 six times. Thus, 5 is a factor of 30. Likewise, the numbers 2, 3 and 4 can all divide 12. Therefore, 2, 3 and 4 are all factors of 12. Let's limit our discussion to integers only.

The **greatest common factor (GCF)** or greatest common divisor (GCD) of two integers is the *largest* number that can evenly divide both of those integers. The GCF is sometimes also referred to as the highest common factor (or divisor). To determine the GCF of two (or more) integers, write lists of all the factors of those integers, and pick the largest integer common to those lists.

1.057 Find the GCF of 24 and 36.

First, systematically list all the possible ways two integers can be multiplied to yield 24:
1×24
2×12
3×8
4×6

Thus, the factors of 24 are:
1, 2, 3, 4, 6, 8, 12, 24

Next, do a similar exercise for 36:
1×36
2×18
3×12
4×9
6×6

The factors of 36 are:
1, 2, 3, 4, 6, 9, 12, 18, 36

What appears to be the *greatest* integer common to both the lists? It's 12. Therefore, the GCF of 24 and 36 is 12.

1.058 What is the GCF of 35 and 70?

The possible ways to obtain 35 as a product are:
1×35
5×7

The factors of 35 are:
1, 5, 7, 35

Similarly, 70 can be obtained in the following ways:
1×70
2×35
5×14
7×10

The factors of 70 are:
1, 2, 5, 7, 10, 14, 35, 70

The greatest integer on both lists is 35. So the GCD of 35 and 70 is 35 itself!

PRACTICE PROBLEMS

In the next four problems, find the GCF of the given integers.

1.059 20 and 12   **1.060** 45 and 72   **1.061** 16, 40 and 12   **1.062** 35, 45 and 55

LOWEST COMMON MULTIPLE (LCM)

We get a **multiple** when we *multiply* a number with another number. We can systematically list the multiples of any number in increasing order. For example, the multiples of 2 are: 2, 4, 6, 8, 10, 12, ... Likewise, the multiples of 3 are 3, 6, 9, 12, 15, 18, ... What is the *lowest* number common to both of these lists? It's 6. Therefore, the **lowest common multiple (LCM)** or least common multiple of 2 and 3 is 6. Let's solve a few examples to reinforce what we've just learned.

1.063 Find the LCM of 6 and 8.

Multiples of 6 are: 6, 12, 18, 24, 30, ...
Multiples of 8 are: 8, 16, 24, 32, 40, ...

What is the *lowest* number that appears on both the lists? It's 24. Therefore, the LCM of 6 and 8 is 24.

1.064 What is the LCM of 16 and 10?

Multiples of 16 are: 16, 32, 48, 64, 80, 96, …
Multiples of 10 are 10, 20, 30, 40, 50, 60, 70, 80, …

The number 80 is the *lowest* number common to both lists, and is the LCM of 16 and 10.

PRACTICE PROBLEMS
In the next four problems, find the LCM of the given integers.

1.065 15 and 25   **1.066** 12 and 16   **1.067** 20, 30 and 40   **1.068** 3, 9 and 27

FRACTIONS

Fractions can be of three types: **proper**, **improper** and **mixed**. Refer to page 1 for an introduction on each type. Before we study arithmetic operations on fractions, let's first understand the conversions back and forth between improper fractions and mixed fractions, best clarified with solved examples.

CONVERTING IMPROPER FRACTIONS INTO MIXED FRACTIONS

1.069 Convert $\dfrac{35}{4}$ into a mixed fraction.

Recall that $\dfrac{35}{4}$ is the same as $35 \div 4$. So, carry out the actual division. 4 times 8 gives 32. That's as close to 35 as you can get (without exceeding it). So the answer would be 8. But wait! You need 3 more to reach 35. Meaning, the remainder is 3. So the answer is 8 and $\dfrac{3}{4}$, written as $8\dfrac{3}{4}$.

1.070 Convert $\dfrac{12}{7}$ into a mixed fraction.

7 times 1 gives 7, the remainder being 5. Thus, the answer is $1\dfrac{5}{7}$.

1.071 Convert $\dfrac{13}{5}$ into a mixed fraction.

$5 \times 2 = 10$, the remainder being 3. Thus, the answer is $2\dfrac{3}{5}$.

Notice that the denominator always remains the same when improper fractions are converted into mixed fractions.

PRACTICE PROBLEMS
Convert the following improper fractions into mixed fractions.

1.072 $\dfrac{25}{7}$ **1.073** $\dfrac{47}{20}$ **1.074** $\dfrac{9}{2}$ **1.075** $\dfrac{16}{3}$

CONVERTING MIXED FRACTIONS INTO IMPROPER FRACTIONS

1.076 Convert $3\dfrac{1}{2}$ into an improper fraction.

Multiply the denominator of the mixed fraction with the outer whole number, and then add the numerator. This gives the numerator of the newly formed improper fraction. The denominator remains the same as before. The illustration below clarifies a seemingly confusing procedure.

2×3 gives 6
$6 + 1$ gives 7

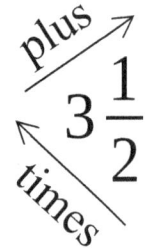

So, the numerator is 7; the denominator remains 2.

The answer is $\dfrac{7}{2}$.

1.077 Convert $11\dfrac{4}{5}$ into an improper fraction.

5×11 gives 55
$55 + 4$ gives 59

So, the numerator is 59; the denominator remains 5.

The answer is $\dfrac{59}{5}$.

PRACTICE PROBLEMS
Convert the following mixed fractions into improper fractions.

1.078 $7\dfrac{2}{3}$ **1.079** $5\dfrac{1}{4}$ **1.080** $15\dfrac{1}{2}$ **1.081** $6\dfrac{3}{8}$

ADDITION/SUBTRACTION OF FRACTIONS

When adding or subtracting fractions, first check their denominators. If the denominators are already the same, then add or subtract the numerators while leaving the denominator unchanged. If the denominators aren't the same, then *raise each fraction to higher terms* such that the new denominator of each fraction is the LCM of the two original denominators. (To review LCM, go to page 8.) Raising a fraction to higher terms simply means multiplying the numerator as well as the denominator of that fraction by a number greater than 1. Once denominators are equalized, add or subtract the numerators.

1.082 $\dfrac{2}{6}+\dfrac{1}{6}$

Notice that the denominators are already the same. Add the numerators.

$$\frac{2+1}{6}$$

$$\frac{3}{6}$$

Now reduce the fraction to its *lowest terms*. Reducing to lowest terms means dividing the numerator as well as the denominator by the same number to slash them down as much as possible. In this case, divide both numbers by 3.

$$\frac{1}{2}$$

1.083 $\dfrac{16}{11}-\dfrac{24}{11}$

The denominators are already the same. Subtract the numerators.

$$\frac{16-24}{11}$$

$$\frac{-8}{11}$$

The fraction is already in its *lowest terms*.

As a side note, there's no difference between $\dfrac{-8}{11}$ or $\dfrac{8}{-11}$ or $-\dfrac{8}{11}$.

1.084 $\dfrac{3}{4}+\dfrac{1}{6}$

The denominators aren't the same. However, their LCM is 12. So we need to raise each fraction to *higher terms* so that their common denominator becomes 12. Multiply the first fraction by 3 at the top and bottom, and the second fraction by 2 at the top and bottom.

$$\frac{3\times3}{4\times3}+\frac{1\times2}{6\times2}$$

$$\frac{9}{12}+\frac{2}{12}$$

$$\frac{9+2}{12}$$

$$\frac{11}{12}$$

1.085 $-\dfrac{8}{9}+\dfrac{5}{7}$

Denominators aren't the same, their LCM being 63. Multiply the first fraction by 7 at the top and bottom, and the second fraction by 9 at the top and bottom.

$$-\frac{8\times7}{9\times7}+\frac{5\times9}{7\times9}$$

$$-\frac{56}{63}+\frac{45}{63}$$

$$\frac{-56+45}{63}$$

$$\frac{-11}{63}$$

1.086 $4\dfrac{1}{2}+1\dfrac{2}{3}$

When adding or subtracting mixed fractions, I think it's easier to first convert them into improper fractions. Visit the previous page for a refresher on how to do that.

$$\frac{2\times4+1}{2}+\frac{3\times1+2}{3}$$

$$\frac{9}{2}+\frac{5}{3}$$

Now equalize the denominators by raising the fractions to higher terms, the LCM of denominators being 6.

$$\frac{9\times3}{2\times3}+\frac{5\times2}{3\times2}$$

$$\frac{27}{6}+\frac{10}{6}$$

$$\frac{27+10}{6}$$

$$\frac{37}{6}$$

Optionally, convert back into a mixed fraction.

$$6\frac{1}{6}$$

1.087 $7\frac{1}{5} - 6\frac{3}{4}$

First, convert both fractions into improper fractions.

$$\frac{5\times7+1}{5} - \frac{4\times6+3}{4}$$

$$\frac{36}{5} - \frac{27}{4}$$

Now equalize their denominators, their LCM being 20.

$$\frac{36\times4}{5\times4} - \frac{27\times5}{4\times5}$$

$$\frac{144}{20} - \frac{135}{20}$$

$$\frac{9}{20}$$

PRACTICE PROBLEMS

1.088 $\frac{9}{7} + \frac{13}{7}$

1.089 $\frac{5}{8} - \frac{3}{8}$

1.090 $\frac{16}{21} + \frac{2}{21} - \frac{11}{21}$

1.091 $\frac{3}{7} + \frac{1}{4}$

1.092 $\frac{1}{2} - \frac{1}{3} - \frac{1}{4}$

1.093 $2\frac{1}{8} - 1\frac{3}{4}$

MULTIPLICATION OF FRACTIONS

When two or more fractions are being multiplied, first try to reduce each fraction to its lowest terms. If the fractions are already in their lowest terms, then try to cross reduce: reduce any numerator with any denominator. Please note that cross reductions are okay only when fractions are being multiplied, not when fractions are being added, subtracted or divided. Once the reductions are done, then multiply the numerators to write the new numerator, and multiply the denominators to write the new denominator.

1.094 $\frac{2}{3} \times \frac{1}{5}$

Notice that neither fraction is reducible, nor can they be cross reduced. So multiply straight across to get the answer.

$$\frac{2\times1}{3\times5}$$

$$\frac{2}{15}$$

1.095 $\frac{3}{5} \times \frac{1}{33} \times 10$

None of the fractions can be reduced, but they can be cross reduced. Also, 10 means $\frac{10}{1}$. We see that 3 goes into 33 eleven times, and 5 goes into 10 two times.

$$\frac{\cancel{3}^1}{\cancel{5}^1} \times \frac{1}{\cancel{33}^{11}} \times \frac{\cancel{10}^2}{1}$$

Now multiply straight across.

$$\frac{1\times1\times2}{1\times11\times1}$$

$$\frac{2}{11}$$

1.096 $\frac{7}{12} \times \frac{5}{14} \times \frac{24}{25}$

None of these fractions can be reduced, but several cross reductions are possible: 7 with 14, 5 with 25, and 12 with 24.

$$\frac{\cancel{7}^1}{\cancel{12}^1} \times \frac{\cancel{5}^1}{\cancel{14}^2} \times \frac{\cancel{24}^2}{\cancel{25}^5}$$

$$\frac{1\times1\times2}{1\times2\times5}$$

$$\frac{2}{10}$$

$$\frac{1}{5} \quad \text{in its } \textit{lowest terms}$$

PRACTICE PROBLEMS

1.097 $\dfrac{6}{7} \times \dfrac{2}{9}$

1.098 $\dfrac{3}{8} \times 5 \times \dfrac{1}{8}$

1.099 $4\dfrac{1}{2} \times \dfrac{21}{25} \times \dfrac{2}{7}$

DIVISION OF FRACTIONS

Dividing one fraction by another is the exact same as multiplying the first fraction by the *reciprocal of the second fraction*. Reciprocal of a fraction, in simple language, means a fraction turned upside down. If a mixed fraction is the one dividing another fraction, then first convert it into an improper fraction before taking its reciprocal. As always, when multiplying two fractions, seek to reduce the terms whenever possible.

1.100 $\dfrac{2}{5} \div \dfrac{8}{15}$

Multiply the first fraction by the reciprocal of the second fraction.

$$\dfrac{2}{5} \times \dfrac{15}{8}$$

Neither fraction can be reduced, but cross reductions are possible: 2 with 8, and 5 with 15.

$$\dfrac{\cancel{2}^{1}}{\cancel{5}^{1}} \times \dfrac{\cancel{15}^{3}}{\cancel{8}^{4}}$$

Multiply straight across.

$$\dfrac{1 \times 3}{1 \times 4}$$

$$\dfrac{3}{4}$$

1.101 $\dfrac{\;7/11\;}{28/55}$

First fraction times the reciprocal of the second fraction.

$$\dfrac{7}{11} \times \dfrac{55}{28}$$

Neither fraction can be reduced, but cross reductions are possible: 7 with 28, and 11 with 55.

$$\dfrac{\cancel{7}^{1}}{\cancel{11}^{1}} \times \dfrac{\cancel{55}^{5}}{\cancel{28}^{4}}$$

Multiply straight across.

$$\dfrac{1 \times 5}{1 \times 4}$$

$$\dfrac{5}{4}$$

$$1\dfrac{1}{4}$$

1.102 $\dfrac{\dfrac{25}{26}}{1\dfrac{2}{13}}$

Rewrite the mixed fraction as an improper fraction.

$$\dfrac{\dfrac{25}{26}}{\dfrac{13 \times 1 + 2}{13}}$$

$$\dfrac{\dfrac{25}{26}}{\dfrac{15}{13}}$$

First fraction times the reciprocal of the second fraction.

$$\dfrac{25}{26} \times \dfrac{13}{15}$$

Neither fraction can be reduced, but cross reductions are possible: 25 with 15, and 26 with 13.

$$\dfrac{\cancel{25}^{5}}{\cancel{26}^{2}} \times \dfrac{\cancel{13}^{1}}{\cancel{15}^{3}}$$

Multiply straight across.

$$\dfrac{5 \times 1}{2 \times 3}$$

$$\dfrac{5}{6}$$

PRACTICE PROBLEMS

1.103 $\dfrac{\dfrac{9}{8}}{\dfrac{3}{32}}$

1.104 $\dfrac{16}{81} \div \dfrac{8}{9}$

1.105 $\dfrac{1\dfrac{23}{40}}{\dfrac{7}{25}}$

DECIMALS

A **decimal** (noun form) is essentially a fraction, only expressed differently. For example, 0.5 is really $\dfrac{1}{2}$. The 0 is the whole number component, and the .5 is the fractional component. The dot (.) is called the **decimal point**. The 5 is called the **decimal digit**. To convert a decimal into a fraction, look at the decimal digits. Place those digits over a power of 10 (meaning 10 or 100 or 1000 etc.) that has *as many zeroes as there are decimal digits*. So, 0.5 becomes $\dfrac{5}{10}$ which can be reduced to $\dfrac{1}{2}$.

Consider another example, let's say, −0.25. How many decimal digits does it have? Two. So it'd be $-\dfrac{25}{100}$ which can be reduced to $-\dfrac{1}{4}$. Yet another example: 0.8000. It has four decimal digits. So it'd be $\dfrac{8,000}{10,000}$ which can be reduced to $\dfrac{8}{10}$ which can further be reduced to $\dfrac{4}{5}$. See the point?

What if there's a whole number component that's *not zero*? For example, 3.14. In such cases, *keep the whole number component unchanged,* and write the decimal digits in the form of a fraction. So, 3.14 would become $3\dfrac{14}{100}$. The fractional component $\dfrac{14}{100}$ can be reduced to $\dfrac{7}{50}$. So the answer is $3\dfrac{7}{50}$ which is a mixed fraction.

1.106 Rewrite 4.75 as a fraction.

There are *two* decimal digits. Place them over 100 (because 100 has *two* zeroes). Keep the whole number component the same.

$4\dfrac{75}{100}$

Reduce the fractional component.

$4\dfrac{3}{4}$

1.107 Rewrite −0.4 as a fraction.

There is *one* decimal digit. Place it over 10 (because 10 has *one* zero).

$-\dfrac{4}{10}$

Reduce the fraction.

$-\dfrac{2}{5}$

1.108 Rewrite 21.019 as a fraction.

There are *three* decimal digits. Place them over 1000 (because 1000 has *three* zeroes). Keep the whole number component the same.

$21\dfrac{019}{1000}$

It's not a good practice to write 019 in the numerator. Instead, write 19.

$21\dfrac{19}{1000}$

It's not reducible, and is the answer.

PRACTICE PROBLEMS

Rewrite the following decimals as fractions.

1.109 2.25   **1.110** −0.48   **1.111** 61.006   **1.112** −9.5

PERCENTS

A **percent** is yet another way of expressing a fraction. The word "percent" is derived from the Latin phrase *per centum* which means "per one hundred." The percentage symbol (%) appears after the number. So, 55% means 55 out of a hundred, or more specifically $\frac{55}{100}$ which is a fraction and can be reduced to $\frac{11}{20}$. Alternatively, $\frac{55}{100}$ can also be written as a decimal. To do that, it's important to understand that 55 essentially means 55.0 which when divided by 100 gives 0.55 because all you have to do is shift the decimal point *two digits to the left*. The general rule here is that when a number is *divided* by a power of 10 (10 or 100 or 1000 etc.), the decimal point in the number shifts to the *left* by a number of digits equal to the number of zeroes in the power of 10.

The following examples demonstrate the idea.

➢ $\frac{125.4}{100} = 1.254$ Two zeroes in 100, so decimal point moved two digits to the *left*.

➢ $\frac{7}{1000} = \frac{7.0}{1000} = \frac{0000007.0}{1000} = 0000.0070 = 0.007$ Extra zeroes were added (doesn't matter how many; just plenty) before 7 to make room for the decimal point to move to the *left*. The excess ones were later discarded. Also, the zero after 7 doesn't matter once the decimal point is placed. So it was discarded as well.

➢ $\frac{-356.217}{10000} = -\frac{000356.217}{10000} = -00.0356217 = -0.0356217$

➢ $\frac{54321}{100} = \frac{54321.0}{100} = 543.210 = 543.21$

I hope you get the point. In case of percents, you'll always need to move the decimal point *exactly two digits to the left*. That makes the task relatively easy, doesn't it?

1.113 Write 27% as a decimal.

$$27\% = \frac{27}{100} = \frac{27.0}{100} = 0.270 = 0.27$$

1.114 Convert −8% into a decimal.

$$-8\% = -\frac{8}{100} = -\frac{8.0}{100} = -\frac{00008.0}{100} = -000.080 = -0.08$$

1.115 Rewrite 25.6% as a decimal.

$$25.6\% = \frac{25.6}{100} = 0.256$$

1.116 Convert 130% into a decimal.

$$130\% = \frac{130}{100} = \frac{130.0}{100} = 1.300 = 1.3$$

PRACTICE PROBLEMS
Rewrite the following percents as decimals.

1.117 −28.8% **1.118** 4% **1.119** 1.09% **1.120** 300%

Rewrite the following percents as fractions reduced to lowest terms where possible. <u>Do not convert further into decimals</u>.

1.121 −25% **1.122** 53% **1.123** −10% **1.124** 145%

CONVERTING DECIMALS INTO PERCENTS

To convert a decimal into a percent, multiply it by 100. When doing so, shift the decimal point *two digits to the right*. Then attach the % symbol and you're done! The general rule when *multiplying* a decimal by a power of 10 (10 or 100 or 1000) is to shift the decimal point to the *right* by a number of spaces equal to the number of zeroes in the power of 10.

The following examples demonstrate the idea.

➢ $9.46728 \times 1000 = 9467.28$

➢ $0.003487 \times 10000 = 00034.87 = 34.87$ After shifting the decimal point, the unnecessary zeroes were removed.

➢ $-365 \times 100 = 365.0 \times 100 = 365.00000 \times 100 = 36500.000 = 36500$ Extra zeroes were added (doesn't matter how many; just plenty) after the decimal point to make room for the decimal point to move to the *right*. The excess ones were later discarded.

➢ $-0.0005137 \times 100 = -000.05137 = -0.05137$

I hope you get the point. In case of converting decimals to percents, you'll always need to move the decimal point *exactly two digits to the right*. At the end, don't forget to attach the % symbol!

1.125 Convert 0.87 into a percent.
$$0.87 \times 100 = 87.0 = 87\%$$

1.127 What's the percent equivalent of 1.14?
$$1.14 \times 100 = 114.0 = 114\%$$

1.126 Rewrite −0.065 as a percent.
$$-0.065 \times 100 = -006.5 = -6.5\%$$

1.128 Write 9 as a percent.
$$9 \times 100 = 900\%$$

PRACTICE PROBLEMS
Convert the following into percents.

1.129 −0.971 **1.130** 3 **1.131** 0.0153 **1.132** 6.28

CONVERTING FRACTIONS INTO DECIMALS

This is the simplest of all conversions. You'd only have to calculate the value of the fraction by actually dividing the numerator by the denominator. Teaching you *how to divide two integers* is beyond the scope of this book.

1.133 Convert $\dfrac{3}{5}$ into a decimal.

$$\frac{3}{5} = 3 \div 5 = 0.6$$

1.134 Rewrite $\dfrac{17}{3}$ as a decimal.

$$\frac{17}{3} = 17 \div 3 = 5.666\ldots = 5.\overline{6} \quad \text{The recurring digit 6 is written with a bar above it.}$$

1.135 Find the decimal equivalent of $-4\dfrac{1}{5}$.

$$-4\frac{1}{5} = -\left(4+\frac{1}{5}\right) = -(4+(1 \div 5)) = -(4+0.2) = -4.2$$

1.136 Write $10\dfrac{1}{6}$ in decimal form.

$$10\frac{1}{6} = 10+\frac{1}{6} = 10+(1 \div 6) = 10+0.16666\ldots = 10+0.1\overline{6} = 10.1\overline{6}$$

PRACTICE PROBLEMS
Convert the following into decimals.

1.137 $1\dfrac{8}{9}$ **1.138** $\dfrac{4}{7}$ **1.139** $-15\dfrac{5}{6}$ **1.140** $-3\dfrac{1}{4}$

CONVERTING FRACTIONS INTO PERCENTS

First find the decimal equivalent of the fraction. Then multiply it by 100, essentially moving the decimal point two digits to the right. Finally attach the % symbol.

1.141 Convert $-\dfrac{3}{8}$ into a percent.

$$-\frac{3}{8} \times 100 = -0.375 \times 100 = -037.5 = -37.5\%$$

1.142 Rewrite $2\frac{1}{3}$ as a percent.

$$2\frac{1}{3}\times100 \;=\; \left(2+\frac{1}{3}\right)\times100 \;=\; (2+0.333....)\times100 \;=\; 2.3333...\times100 \;=\; 233.33...=233.\overline{3}\,\%$$

1.143 Find the percent equivalent of $\frac{11}{7}$.

$$\frac{11}{7}\times100 \;=\; 1.571428571428....\times100 \;=\; 157.142857142857.... \;=\; 157.\overline{142857}\,\%$$

PRACTICE PROBLEMS
Convert the following into percents.

1.144 $-\dfrac{4}{5}$ **1.145** $4\dfrac{2}{9}$ **1.146** $\dfrac{11}{6}$

RADICALS

If I asked you what is 4^2 (meaning, *four squared*), you would (probably look back at page 5 and) tell me it is 4×4 which is 16. But if I asked you what number *squared* gives 16? Your answer would then be 4. Meaning, just the way 16 is the *square* of 4, 4 is the *square root* of 16. Please note that -4×-4 also gives 16, so that would make -4 also the square root of 16. But our discussion will be limited to the positive square root only. Square root of 4 is written as $\sqrt{4}$ where 4 is called the **radicand** and $\sqrt{}$ is the square root sign. It is collectively called a **radical** (noun form).

So, what's the square root of 81? It's 9 because $9 \times 9 = 81$. And $\sqrt{49}$? That would be 7. What about $\sqrt{50}$? Now we have a problem. No *integer* when multiplied by itself gives 50. But we can still manipulate it and write it in the form of a *smaller radicand*. Here's how:

Break down 50 as the product of two or more integers until you can't break it down any further.

$\sqrt{50}$

$\sqrt{10\times5}$

$\sqrt{5\times2\times5}$

$\sqrt{5\times5\times2}$ Just rearranged the numbers; no big deal.

Now see if you can form a perfect square while still underneath the square root sign. Yes! $5 \times 5 = 25$ which is a perfect square.

$\sqrt{25\times2}$

The square root of 25 is 5, whereas the square root of 2 doesn't exist as an integer. So $\sqrt{25}$ would now be written outside as 5, while $\sqrt{2}$ would stay unchanged.

$5\sqrt{2}$

This is how you express the square root of a non-integer square in terms of a *smaller radicand*. Please note that $5\sqrt{2}$ means $5\times\sqrt{2}$, not $5+\sqrt{2}$. Also, the 5 on the outside is called the **coefficient**.

As a side note, not all non-integer squares can be expressed in terms of smaller radicands, for example $\sqrt{13}$, $\sqrt{11}$, $\sqrt{30}$ etc.

1.147 Express $\sqrt{24}$ in terms of a smaller radicand.

$\sqrt{6 \times 4}$

$\sqrt{3 \times 2 \times 2 \times 2}$

Now reconstruct a perfect square using a combination of those numbers. We see that $2 \times 2 = 4$.

$\sqrt{3 \times 2 \times 4}$

The square root of 4 is 2. So $\sqrt{4}$ would now be written outside as the coefficient 2.

$2\sqrt{3 \times 2}$

$2\sqrt{6}$

1.148 Rewrite $\sqrt{600}$ with a smaller radicand.

$\sqrt{6 \times 100}$

Notice that 100 is already 10 squared. Look for such perfect squares when writing the product. That would save you some time. $\sqrt{100}$ is 10, so that would come outside as the coefficient.

$10\sqrt{6}$

$10\sqrt{3 \times 2}$

Neither 3 nor 2 is a perfect square. So the answer is $10\sqrt{6}$.

1.149 Write $\sqrt{70}$ with a smaller radicand.

$\sqrt{7 \times 10}$

$\sqrt{7 \times 5 \times 2}$

None of these numbers is a perfect square. So $\sqrt{70}$ can't be written with a smaller radicand.

PRACTICE PROBLEMS
Rewrite the following with smaller radicands.

1.150 $\sqrt{75}$ **1.151** $\sqrt{35}$ **1.152** $\sqrt{32}$

ADDITION/SUBTRACTION OF RADICALS

When adding or subtracting radicals, first make sure the *radicands* are identical. Then add or subtract the coefficients (the numbers outside), while leaving the radicands unchanged. That's all!

1.153 $2\sqrt{3} + 8\sqrt{3}$

Both the radicands are 3. Add the coefficients 2 and 8, and leave the $\sqrt{3}$ unchanged.

$(2+8)\sqrt{3}$

$10\sqrt{3}$

1.154 $5\sqrt{7} - 11\sqrt{7} + 20\sqrt{7}$

The radicands are identical. So perform addition/subtraction of the coefficients and leave the $\sqrt{7}$ unchanged.

$(5-11+20)\sqrt{7}$

$14\sqrt{7}$

1.155 $6\sqrt{3}+\sqrt{12}$

The radicands are not the same. So, first simplify $\sqrt{12}$.
$6\sqrt{3}+\sqrt{4\times3}$
The square root of 4 is 2. So $\sqrt{4}$ would now be written outside as 2, the coefficient.
$6\sqrt{3}+2\sqrt{3}$
Now the radicands are the same. Add the coefficients 6 and 2. Leave the $\sqrt{3}$ unchanged.
$8\sqrt{3}$

1.156 $9\sqrt{20}-\sqrt{125}$

The radicands are not the same. So, first simplify them both.
$9\sqrt{4\times5}-\sqrt{25\times5}$
$\sqrt{4}$ is 2, and $\sqrt{25}$ is 5. Extract them both.
$9\times2\sqrt{5}-5\sqrt{5}$
$18\sqrt{5}-5\sqrt{5}$
Now the radicands are the same. Subtract the coefficient 5 from 18. Leave the $\sqrt{5}$ unchanged.
$13\sqrt{5}$

PRACTICE PROBLEMS

1.157 $5\sqrt{7}-\sqrt{28}$

1.158 $\sqrt{54}-\sqrt{24}+3\sqrt{6}$

1.159 $3\sqrt{50}-4\sqrt{18}$

1.160 $-7\sqrt{27}-5\sqrt{12}$

MULTIPLICATION OF RADICALS

When multiplying two radicals, multiply the coefficients to get the new coefficient, and multiply the radicands to get the new radicand. Simplify the radicand whenever possible.

1.161 $2\sqrt{5}\times3\sqrt{7}$

Multiply the coefficients separately. Multiply the radicals separately.
$2\times3\times\sqrt{5}\times\sqrt{7}$
When two radicals are *multiplied*, their radicands combine into a single radicand product under one radical sign, as demonstrated in the next step below. Note that this doesn't happen when two radicals are *added/subtracted*.
$2\times3\times\sqrt{5\times7}$
$6\sqrt{35}$
The radicand can't be simplified any further. Therefore, this is the answer.

1.162 $-6\sqrt{5}\times2\sqrt{45}$

Multiply the coefficients, and combine the radicands under one radical sign.
$-6\times2\sqrt{5\times45}$
$-12\sqrt{5\times5\times9}$
$-12\sqrt{25\times9}$
You might notice that this is just an exercise in simplifying and rearranging the radicand.
$-12\times5\times3$
No more radicands! Frankly, I don't like the sound of that word. *Radicand?* Meh!
-180

1.163 $3\sqrt{50}\times9\sqrt{2}$

Multiply the coefficients, and combine the radicands under one radical sign.

$3\times9\sqrt{50\times2}$

$27\sqrt{100}$

27×10

270

PRACTICE PROBLEMS

1.164 $2\sqrt{18}\times18\sqrt{2}$ **1.165** $-4\sqrt{10}\times(-3)\sqrt{2}$ **1.166** $\sqrt{200}\times3\sqrt{8}$

DIVISION OF RADICALS

When dividing two radicals, divide (or reduce) the coefficients to get the new coefficient, and combine and reduce the radicands to get the new radicand.

1.167 $\dfrac{2\sqrt{7}}{10\sqrt{14}}$

Before reducing the coefficients as well as the radicands, write them as a product of two separate fractions.

$$\frac{2}{10}\times\frac{\sqrt{7}}{\sqrt{14}}$$

When two radicals are *divided*, their radicands combine into a single radicand fraction under one radical sign, as demonstrated in the next step below.

$$\frac{2}{10}\times\sqrt{\frac{7}{14}}$$

Now reduce both fractions.

$$\frac{1}{5}\times\sqrt{\frac{1}{2}}$$

Now break the radical again into two separate radicals. Then simplify.

$$\frac{1}{5}\times\frac{\sqrt{1}}{\sqrt{2}}$$

$$\frac{1}{5}\times\frac{1}{\sqrt{2}}$$

Now multiply straight across.

$$\frac{1}{5\sqrt{2}}$$

1.168 $\dfrac{\sqrt{54}}{-12\sqrt{6}}$

Remember, the coefficient of $\sqrt{54}$ is 1. It's not *nothing*!

$$\frac{1\sqrt{54}}{-12\sqrt{6}}$$

Write as a product of two separate fractions: one containing the coefficients and the other containing the radicals.

$$\frac{1}{-12}\times\frac{\sqrt{54}}{\sqrt{6}}$$

Now combine the radicands.

$$\frac{1}{-12}\times\sqrt{\frac{54}{6}}$$

$$\frac{1}{-12}\times\sqrt{9}$$

$$\frac{1}{-12}\times3$$

$$\frac{1}{-12}\times\frac{3}{1}$$

$$\frac{1\times3}{-12\times1}$$

$$\frac{3}{-12}$$

$$\frac{1}{-4}$$

$$-\frac{1}{4}$$

1.169 $45 \div 3\sqrt{50}$

$$\frac{45}{3\sqrt{50}}$$

The 3 can readily reduce the 45.

$$\frac{15}{\sqrt{50}}$$

$$\frac{15}{\sqrt{25 \times 2}}$$

$$\frac{15}{5\sqrt{2}}$$

The 5 can readily reduce the 15.

$$\frac{3}{\sqrt{2}}$$

PRACTICE PROBLEMS

1.170 $2\sqrt{8} \div \sqrt{32}$

1.171 $\dfrac{\sqrt{44}}{\sqrt{99}}$

1.172 $3\sqrt{75} \div 5\sqrt{27}$

CHAPTER EXERCISES

1.173 $-21 + (-13) - (+6) - (-2)$

1.174 $-4 \times (-5) \times 2$

1.175 $\dfrac{-16}{-4}$

1.176 $39 \div (-13) \times 8$

1.177 $\left(\dfrac{6 - (-4)}{5}\right)^2 + 6 \times (-2)$

1.178 $-11 + (-8) \div 2 \times (-3)$

1.179 Find the GCF of 36 and 20.

1.180 Find the LCM of 12 and 15.

1.181 Convert $\dfrac{17}{3}$ into a mixed fraction.

1.182 Convert $9\dfrac{2}{5}$ into an improper fraction.

1.183 $\dfrac{1}{3} - \dfrac{5}{4} + \dfrac{7}{8}$

1.184 $2\dfrac{5}{6} + 10\dfrac{1}{3} - 6\dfrac{1}{2}$

1.185 $\dfrac{27}{34} \times 5\dfrac{2}{3}$

1.186 $7\dfrac{1}{6} \div \dfrac{22}{3}$

1.187 Rewrite -0.84 in the form of a fraction and reduce to lowest terms.

1.188 Rewrite 28% first as a fraction. <u>Do not reduce to lowest terms</u>. Then rewrite the fraction as a decimal.

1.189 Convert $\dfrac{5}{4}$ first into a decimal. Then convert the decimal into a percent.

1.190 Convert 65% into a fraction and reduce to lowest terms.

1.191 Convert $\dfrac{11}{12}$ into a percent.

1.192 Rewrite $\sqrt{84}$ in terms of a smaller radicand.

1.193 Can $\sqrt{26}$ be expressed with a smaller radicand?

1.194 $4\sqrt{12} - 11\sqrt{3} + 2\sqrt{75}$

1.195 $-3\sqrt{72} \times (-2)\sqrt{5}$

1.196 $-6\sqrt{44} \div (-5)\sqrt{18}$

CHAPTER 2
POLYNOMIALS AND EXPRESSIONS

The real *algebra* begins in this chapter, where we deal with letters from the alphabet. A letter (x, y, z etc.) used to represent a number in algebra is known as a **variable**, because its value *varies* from example to example. As compared with a variable, a **constant** (5, −2.68, ¼ etc.) is simply a number. It's called a constant because its value remains *constant* (unchanging).

MONOMIALS, POLYNOMIALS AND EXPRESSIONS

The basic building block of algebra is called a **monomial**, comprising a constant, or a variable, or a constant and a variable *multiplied* with each other. For example, 8 is a monomial. So is $5x$, where 5 is the constant and x is the variable. Just y would also be a monomial since here the constant is 1 (it's not *nothing*) and the variable is y. Constants that are multiplied with variables are called **coefficients**. In the monomial $6a^3$, 6 is the coefficient, a is the variable, and 3 is the **degree** of the term.

When two monomials are added or subtracted, we get a **binomial**, for example $2x^3 - 7$. When three monomials are added or subtracted, we get a **trinomial**, as in $3a^2 - 11a + 4$. In general, two or more monomials when added or subtracted comprise what's called a **polynomial**.

Please note that a polynomial can only involve *one variable*. Thus, $16x^2 + 3x - 4$ would be a polynomial, but $5x + 8y$ would not because it involves *two variables* (x and y). Another thing to remember is that the degree of each term in a polynomial can only be zero or a positive *integer*. So that means $x^{-2} + 3x^{1/5} - 21$ isn't a polynomial because it has degrees that are negative and fraction, neither of which is acceptable. So if it can't be classified as a polynomial, then what *can* it be classified as? That's a good question, reader. It's classified as an **expression**. Other examples of expressions would be $6bc + 2c$, $x^2 + 3xy + 19y^5$ etc.

As the title of this chapter suggests, we'll be pursuing a discussion of polynomials and expressions together. I haven't written two chapters treating each entity separately. Instead, we'll be going back and forth between the two. In my teaching experience, it has worked better this way.

EVALUATING POLYNOMIALS AND EXPRESSIONS

Consider the expression $5a + 8b$. What if I told you that the value of a is actually 10, and the value of b is 4? You'd then be able to find the value of the expression, wouldn't you?

I'm guessing here's how you would do that: In the place of a, you'd put 10, and in the place of b, you'd put 4. The operation would look as follows.

$5a+8b$
$5\times10 \ + \ 8\times4$
$50+32$
82

In algebra, multiplication can be written in many ways. For example, 16 times h can be written as $16h$ or $16\times h$ or $16\cdot h$ or $(16)(h)$ or $(16)\times(h)$ or $(16)\cdot(h)$ or $16(h)$. They're all correct! I'll be using these notations throughout the book so that you become comfortable seeing them over time.

2.001 Evaluate the expression $c+\dfrac{a^2}{15\,xy}$ for $a=5$, $c=3$, $x=10$, $y=1$.

Put the given values of a, c, x and y in the expression and calculate.

$3+\dfrac{5^2}{15(10)(1)}$

$3+\dfrac{25}{150}$

$3+\dfrac{1}{6}$

$3\dfrac{1}{6}$

2.002 Evaluate the expression $-4\,pq+3\,x$ for $p=-2$, $q=-3$, $x=8$.

Replace the variables with the values given for them.
$-4\cdot(-2)\cdot(-3) \ + \ 3\cdot(8)$
Before you proceed, review *order of operations* from page 6 if you need to.
$-4\times2\times3 \ + \ 3\times8$
$-24+24$
0

2.003 Evaluate the polynomial $6x^2+15x-20$ for $x=-2$.

$6(-2)^2+15(-2)-20$
$6\times4+(-30)-20$
$24-30-20$
$-6-20$
-26

2.004 Evaluate the expression $\dfrac{8GM}{r^3}-17$ for $r=4$, $G=2$, $M=-4$.

$\dfrac{8\cdot2\cdot(-4)}{4^3}-17$

$\dfrac{-64}{64}-17$

$-1-17$

-18

PRACTICE PROBLEMS

2.005 Evaluate $13a+6bc-1$ for
$a=2$, $b=2$, $c=-1$.

2.006 Evaluate $\dfrac{4c+9q}{c-q}$ for $c=0$, $q=8$.

2.007 Evaluate $b^3+\dfrac{4\pi r^3}{3}$ for
$b=2$, $\pi=3$, $r=1$.

2.008 Evaluate n^3+2n^2-n+7 for $n=-3$.

ADDITION/SUBTRACTION OF POLYNOMIALS AND EXPRESSIONS

When adding or subtracting two monomials, keep the variable portion the same, and add/subtract their *coefficients*. For example, what is $4a+11a$? Well, the coefficients are 4 and 11. Add them to get 15. The variable portion, a, remains as is. So the answer would be $15a$. To understand it better, imagine adding 4 apples to 11 apples. What's the result? You'd get 15 apples. Meaning, you'd be adding 4 and 11 *of the same type,* to get 15 of that very type.

Now imagine adding the polynomial $3x^2+8$ to the polynomial $9x^2-5$. When doing so, first notice that there are two *types of terms* in each polynomial. So add the x^2 terms separately, and add the constant terms separately. (When you're adding terms of the same type, you're adding what are called **like terms**. The two x^2 terms are *like terms*. Similarly, the two constant terms are *like terms*.) That would give $3x^2+9x^2$ which is $12x^2$, and $8+(-5)$ which is the same as $8-5$ which is 3. Thus, the answer would be $12x^2+3$.

2.009 Add the following polynomials: $2x^2+5x+3$ and $-3x^2+7x-4$

$$\left(2x^2+5x+3\right) + \left(-3x^2+7x-4\right)$$

First, regroup the x^2 terms separately, the x terms separately, and the constant terms separately. Then do the addition.

$$\overbrace{2x^2+(-3x^2)}\quad\overbrace{+5x+7x}\quad\overbrace{+3+(-4)}$$

The $\overbrace{\text{wave}}$ shown above the text is not a mathematical symbol. It's just my way of highlighting that the terms have been grouped together.

$$\overbrace{2x^2-3x^2}\quad\overbrace{+5x+7x}\quad\overbrace{+3-4}$$

Now simplify the coefficients of the *like terms*.

$$-1x^2+12x+(-1)$$
$$-1x^2+12x-1$$

In algebra, it's not a good practice to write $-1x^2$ in the final answer. Instead, we just write $-x^2$. Meaning, when the coefficient of a monomial is 1 or -1, we just write the variable with or without the negative sign depending on what the answer is.

$$-x^2+12x-1$$

2.010 Subtract $8x^3-13x+x^2-4$ from $2-11x^3+x-6x^2$.

$$\left(2-11x^3+x-6x^2\right)-\left(8x^3-13x+x^2-4\right)$$

First, let's subtract the *like terms* by grouping them.

$$\overbrace{2-(-4)}\quad\overbrace{-11x^3-8x^3}\quad\overbrace{+x-(-13x)}\quad\overbrace{-6x^2-x^2}$$

Before you proceed, simplify the excess signs between the terms. In case you'd like a refresher, visit example 1.003 from page 3 and read page 2 for the rules pertaining to signs.

$$\overbrace{2+4} \quad \overbrace{-11x^3-8x^3} \quad \overbrace{+x+13x} \quad \overbrace{-6x^2-x^2}$$

Remember that x means $1x$, and x^2 means $1x^2$.

$$6-19x^3+14x-7x^2$$

Realistically, this is the answer. But in algebra, it's generally a good practice to present the terms of a polynomial in *reducing order of their degrees*.

$$-19x^3-7x^2+14x+6$$

2.011 $(3x^4+2x+8)+(x^2+5x+6)-(2x^4-x+11)$

This time, first group the *like terms* in reducing order of their degrees.

$$\overbrace{3x^4-2x^4} \quad \overbrace{+x^2} \quad \overbrace{+2x+5x-(-x)} \quad \overbrace{+8+6-11}$$

Notice that the x^2 term has no other term to pair up with. It stands alone. There's no need to get emotional about it. Trust me, x^2 will be just fine! Simplify the excess signs between the terms.

$$\overbrace{3x^4-2x^4} \quad \overbrace{+x^2} \quad \overbrace{+2x+5x+x} \quad \overbrace{+8+6-11}$$

$$x^4+x^2+8x+3$$

2.012 $(5a^2b-3x)+(8a^2b+7x)$

This is an expression, not a polynomial. But the rules for grouping the *like terms* are identical.

$$\overbrace{5a^2b+8a^2b} \quad \overbrace{-3x+7x}$$

$$13a^2b+4x$$

2.013 $(14-12xa+6bc)-(12cb-5ax)$

Please note that xa is the same as ax because the variables a and x are being multiplied with each other one way or the other. Similarly, bc is the same as cb. Group the *like terms*.

$$14 \quad \overbrace{-12xa-(-5ax)} \quad \overbrace{+6bc-12cb}$$

$$14 \quad \overbrace{-12xa+5ax} \quad \overbrace{+6bc-12cb}$$

$$14-7ax-6bc$$

This is an expression, not a polynomial. If you'd like to optionally rearrange the terms and write the 14 at the end, you can. But it's not absolutely necessary.

2.014 $(16-2ab)-(5x^2+13ab+1)-(-4ba+3x^2)$

Group the *like terms*.

$$\overbrace{16-1} \quad \overbrace{-2ab-13ab-(-4ba)} \quad \overbrace{-5x^2-3x^2}$$

$$\overbrace{16-1} \quad \overbrace{-2ab-13ab+4ba} \quad \overbrace{-5x^2-3x^2}$$

$$15-11ab-8x^2$$

2.015 $(5b^2+3)+(2b-b^3)-(b^2-11b+12)$

Observe that this is a polynomial because it has only one variable. Group the *like terms*.

$$-b^3 \quad \overbrace{+5b^2-b^2} \quad \overbrace{+2b-(-11b)} \quad \overbrace{+3-12}$$

$$-b^3 \quad \overbrace{+5b^2-b^2} \quad \overbrace{+2b+11b} \quad \overbrace{+3-12}$$

$$-b^3+4b^2+13b-9$$

PRACTICE PROBLEMS

2.016 $(-a^2+3a-1)+(8a^2+5a-2)$

2.017 $(c^2y+5)-(-2yc^2+12k)+(6-6k)$

2.018 $(7ab+10-b)-(17-2ba+c)-(4ab-6c)$

2.019 $(3p+4q)-(2p+8q)+(5p-6)$

MULTIPLICATION OF MONOMIALS

From the brief discussion we've had on exponents in the last chapter (see page 5), we know that $3\times3=3^2$. Likewise, $3\times3\times3\times3=3^4$. Along these lines, we can conclude that $3\times3^3=3^4$. Similarly, $3^5\times3^{11}=3^{16}$. Do you see the pattern? When multiplying two or more exponential terms with the same base, *the exponents get added up*. In algebra, we can use the same rule when using variables, too: $a^2\times a^6=a^8$ $m^7\times m\times m^7=m^{15}$ We'll pursue a greater discussion on exponents later in this chapter, but for now, this should do.

When two or more monomials are multiplied, the coefficients should be multiplied separately, and the variables should be multiplied separately. The term should then be written as a whole, as is demonstrated in the following examples:

➤ $x^2\cdot x^3=x^{2+3}=x^5$

➤ $a^{10}\cdot a\cdot a^3=a^{10+1+3}=a^{14}$

➤ $2a\times6a=(2\times6)a^{1+1}=12a^2$

➤ $(5x)\cdot(8x^2)=(5\times8)x^{1+2}=40x^3$

➤ $-3(7c)=(-3\times7)c=-21c$

➤ $3b\times8c=(3\times8)bc=24bc$

➤ $(-4t)\times(-2t^2)\times(2v)=[(-4)\times(-2)\times(2)]t^{1+2}v=16t^3v$

Notice that in each instance above, the coefficients multiplied out to form the new coefficient; the variables multiplied out to form the new variable component. In the last but one row, the variables were different, so they simply got written side by side. In the last row, the variables were different but t did get multiplied with t^2, whereas v remained as is. *Multiply variables only with their own type.*

DISTRIBUTING ONE TERM ONTO OTHERS

When a monomial is multiplied to a polynomial (or to an expression), the monomial gets multiplied to each term in the polynomial (or expression), one term at a time. The process is called **distribution**, and the property is called **distributive property**. *Distribution*, although a fancy term, simply means multiplication. Let's see how it works.

2.020 $p(q+r)$

Multiply p with each of the terms q and r within the parentheses.

$\overbrace{p\cdot q}\quad\overbrace{+p\cdot r}$

$pq+pr$

2.021 $\quad -2c(a-3b)$

Multiply $-2c$ with each of the interior terms. Treat the $(-)$ sign in front of $3b$ as a part of $3b$ itself. In general, *the signs in front of the inner terms are a part of the terms.*

$$\overbrace{-2c(a)} \quad \overbrace{-2c(-3b)}$$
$$-2ca+6cb$$

2.022 $\quad 5(x^2+3x+1)$

Distribute 5 onto each of the three inner terms. Distributing just means multiplying.

$$\overbrace{5(x^2)} \quad \overbrace{+5(3x)} \quad \overbrace{+5(1)}$$
$$5x^2+15x+5$$

2.023 $\quad 2a(3a^2+4a-5)$

Distribute $2a$ over the inner terms.

$$\overbrace{2a\times3a^2} \quad \overbrace{+2a\times4a} \quad \overbrace{+2a\times(-5)}$$
$$6a^3+8a^2-10a$$

2.024 $\quad 5x^2(2x-7xy+9y)$

Distribute $5x^2$ onto the inner terms.

$$\overbrace{5x^2\cdot(2x)} \quad \overbrace{+5x^2\cdot(-7xy)} \quad \overbrace{+5x^2\cdot(9y)}$$
$$10x^3-35x^3y+45x^2y$$

PRACTICE PROBLEMS

2.025 $\quad 4(x+3y)$

2.026 $\quad -2(5a+3a^2b)$

2.027 $\quad ab(-a+b^2-2ab)$

2.028 $\quad 3xy(2x-4y+6)$

2.029 $\quad -p^2q(11p-12pq^2)$

2.030 $\quad 4mn(3n^2+5m-mn^2)$

DISTRIBUTING TWO TERMS ONTO OTHERS

This section builds on top of the previous section (distributing one term onto others). If you'd need to review the previous section, please do so before continuing.

When a polynomial (or expression) containing two terms is multiplied with another polynomial (or expression) containing two or more terms, each term from the first polynomial gets a chance to be multiplied with each term of the second one. Consider the following example.

$$(a+b)\cdot(c+d)$$

The first expression has the terms a and b. Pick up the term a and distribute it onto the second set of parentheses. Then come back and pick up the term b and distribute it onto the second set of parentheses. The illustration below shows how.

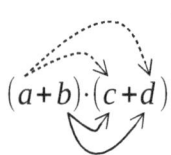

The result will be the following.

$$\overbrace{a \cdot c} \quad \overbrace{+a \cdot d} \quad \overbrace{+b \cdot c} \quad \overbrace{+b \cdot d}$$
$$ac + ad + bc + bd$$

There are no *like terms* here to combine. Therefore, this is the answer.

2.031 $(x+2) \cdot (x+6)$

First distribute x onto $(x+6)$. Then distribute 2 onto $(x+6)$.

$$\overbrace{x \cdot x} \quad \overbrace{+x \cdot 6} \quad \overbrace{+2 \cdot x} \quad \overbrace{+2 \cdot 6}$$
$$x^2 + 6x + 2x + 12$$

Combine the *like terms* $6x$ and $2x$.

$$x^2 + 8x + 12$$

2.032 $(a+2b) \cdot (3a-b)$

Distribute a onto $(3a-b)$. Then distribute $2b$ onto $(3a-b)$.

$$\overbrace{a \cdot (3a)} \quad \overbrace{+a \cdot (-b)} \quad \overbrace{+2b \cdot (3a)} \quad \overbrace{+2b \cdot (-b)}$$
$$3a^2 - ab + 6ba - 2b^2$$

Combine the *like terms* $-ab$ and $6ba$.

$$3a^2 + 5ab - 2b^2$$

2.033 $(2x+5)(2x-5)$

Distribute $2x$ onto $(2x-5)$. Then distribute 5 onto $(2x-5)$.

$$\overbrace{2x(2x)} \quad \overbrace{+2x(-5)} \quad \overbrace{+5(2x)} \quad \overbrace{+5(-5)}$$
$$4x^2 - 10x + 10x - 25$$

Combine the *like terms* $-10x$ and $10x$ (which cancel each other out).

$$4x^2 - 25$$

2.034 $(p+3q)(2m-n)$

Distribute p over $(2m-n)$. Then distribute $3q$ over $(2m-n)$.

$$\overbrace{p(2m)} \quad \overbrace{+p(-n)} \quad \overbrace{+3q(2m)} \quad \overbrace{+3q(-n)}$$
$$2pm - pn + 6qm - 3qn$$

There are no *like terms* here to combine. So, this is the answer.

PRACTICE PROBLEMS

2.035 $(c+6)(c-1)$

2.036 $(4x-3y)(x+2y)$

2.037 $(m+2a)(3m-5)$

2.038 $(4+3p)(4-3p)$

FACTORING POLYNOMIALS AND EXPRESSIONS

In the last two sections, we learned how to distribute one or two terms over several other terms. Now we'll study the process of **factoring**, which is the exact opposite of distributing. In factoring, we extract or *factor out* the greatest common factor (GCF) from each term.

For example, consider the terms $15x$ and $10xy$. What is the GCF of the coefficients 15 and 10? It's 5 because 5 is the greatest number that goes into both 15 and 10. What is the GCF of x and xy? It's x because x is the only variable that goes into both x and xy. Therefore, the GCF of $15x$ and $10xy$ is $5x$.

Now consider the term $15x + 10xy$. We can factor out (or extract) $5x$ from each term and rewrite:
$$5x(3+2y)$$
This is the answer. Observe that if we distribute $5x$ onto $(3+2y)$, we' get back our original $15x + 10xy$. This is how factoring is the exact opposite of distributing.

Another example: What is the GCF of $4a^2$ and $20a^3$? That would be $4a^2$ because 4 is the greatest number that goes into both the coefficients 4 and 20, and a^2 (not just a) is the greatest variable term that goes into both a^2 and a^3.

Now consider the term $4a^2 - 20a^3$. We can factor out $4a^2$ from each term and rewrite:
$$4a^2(1-5a)$$
This is the answer. Understand that when $4a^2$ got factored out of the first term, it left behind 1 inside the parentheses; it didn't leave behind *nothing*. If we distribute $4a^2$ onto $(1-5a)$, we get back our original $4a^2 - 20a^3$. See how factoring and distributing are opposites of each other?

2.039 Factor $3a+3b$.

Notice that 3 is common to both terms and is the GCF. There's no common variable in the two terms. Factor out 3 and rewrite.
$$3(a+b)$$
This is the answer. If you distribute 3 onto a and b, you'd get back the original $3a+3b$.

2.040 Factor $ab+ac$.

a is the only variable common to both terms. Factor it out.
$$a(b+c)$$
Distribution of a onto b and c results in the original expression. This is how you can check if your answer was right.

2.041 Factor $4m+8n$.

The GCF of 4 and 8 is 4. There's no common variable in the two terms. Factor out 4.
$$4(m+2n)$$
Distributing 4 onto m and $2n$ would yield the original expression.

2.042 Factor $-10xy-4y$.

The GCF of 10 and 4 is 2. The GCF of xy and y is y. Pay attention to the signs as well. It appears we can optionally factor out the (–) sign as well. Meaning, factor out $-2y$.
$$-2y(5x+2)$$
To check your answer, distribute and notice you'll get back the original expression.

2.043 Factor $15x^3+5x^2-25x$.

The GCF of 15, 5 and –25 is 5. The GCF of x^3, x^2 and x is just x. So, factor out $5x$.
$$5x(3x^2+x-5)$$

2.044 Factor $16a^2-24a+8$.

The GCF of 16, –24 and 8 is 8. There's no variable in the last term. Meaning, no variable can be factored out. Factor out 8.

$8(2a^2-3a+1)$

Notice that when 8 was factored out of the last term, it left behind 1. It didn't leave behind *nothing*. Distribute to verify that you get back the original expression.

2.045 Factor $4ab+6bc-12abc$.

The GCF of 4, 6 and –12 is 2. The GCF of *ab*, *bc* and *abc* is just *b*. Factor out 2*b*.

$2b(2a+3c-6ac)$

PRACTICE PROBLEMS
Factor the following polynomials/expressions.

2.046 $12a+12b$

2.048 $25c^2+10c-5$

2.050 $8r^2-80r+16r^3$

2.047 $6xy+16xyz$

2.049 $-4x^3-10x^2-22x$

2.051 $-5mn+15m^2n^2-25mn^2$

FACTORING THE QUADRATIC: ax^2+bx+c

A polynomial of the form ax^2+bx+c is called a **quadratic** (noun form) where *a*, *b* and *c* are actual numbers, not just letters. For example, $3x^2+13x+4$ is a quadratic where $a=3$, $b=13$ and $c=4$. *Some* quadratics can be factored, while others cannot. This section deals with the ones that *can* be factored. Let's see how:

$3x^2 + 13x + 4$ **product**

First, find the **product** *ac* (meaning, *a* times *c*). That would be $3\times4=12$.

Next, systematically list the different ways the product *ac*, which is 12, can be obtained.
1×12
2×6
3×4

Which of the above pairs of numbers when *added* gives the **sum** 13 (coefficient of the middle term)?

sum

$3x^2 + 13x + 4$

That would be 1 and 12. Pick that pair and split the middle term as either $1x+12x$ or $12x+1x$ (doesn't matter which way). For the sake of clarity, I suggest you write 1*x*, not just *x*.

$3x^2+1x+12x+4$

Group the first two terms. Group the next two terms. Then factor the most you can from each group.

$3x^2+1x \quad +12x+4$
$1x(3x+1) +4(3x+1)$

Notice that $(3x+1)$ shows up in both the groups. Factor it out. That leaves behind $1x$ from the first group, and $+4$ from the second group. These leftovers, $1x$ and $+4$, form the second term.

$(3x+1)(1x+4)$

That's really $(3x+1)(x+4)$ where $1x$ was changed back to simply x.

2.052 Factor the quadratic x^2+5x+6.

Notice that x^2 really means $1x^2$.

$1x^2+5x+6$

When compared with ax^2+bx+c, we see that $a=1$, $b=5$, $c=6$.

What's the product ac? It is $1 \times 6 = 6$.

The different ways to obtain the product 6 are:

$1 \times 6 \qquad \boxed{2 \times 3} \leftarrow$

The coefficient of middle term, 5, can be obtained by *adding* this pair.

Therefore, split the middle term as $2x+3x$.

$1x^2+2x+3x+6$

Group the first two terms. Group the next two terms. Factor the most you can from each group.

$\overbrace{1x^2+2x} \quad \overbrace{+3x+6}$
$1x(x+2)+3(x+2)$

The term $(x+2)$ shows up in both groups. Factor it out, leaving behind $1x$ and $+3$.

$(x+2)(1x+3)$

Finally, rewrite $1x$ as just x. It was originally changed to $1x$ so that factoring would become easy.

$(x+2)(x+3)$

2.053 Factor the quadratic $p^2-10p+21$.

Notice that the variable is p, not x. Also, p^2 means $1p^2$. Comparing $1p^2-10p+21$ with the quadratic ax^2+bx+c, we see that $a=1$, $b=-10$, $c=21$.

What's the product ac? It is $1 \times 21 = 21$.

The middle term is -10, not just 10. So, consider negative possibilities also when coming up with factors of the product 21. The different ways to obtain the product 21 are:

$1 \times 21 \qquad (-1) \times (-21) \qquad 3 \times 7 \qquad \boxed{(-3) \times (-7)} \leftarrow$

The coefficient of middle term, -10, can be obtained by *adding* this pair.

Therefore, split the middle term as $(-3p)+(-7p)$.

$1p^2+(-3p)+(-7p)+21$

Group the first two terms. Group the next two terms.

$\overbrace{1p^2+(-3p)} \quad \overbrace{+(-7p)+21}$

Simplify the excess signs between the terms. Factor the most you can from each group.

$\overbrace{1p^2-3p} \quad \overbrace{-7p+21}$
$1p(p-3)-7(p-3)$

From the second set of parentheses, I factored out -7 (not $+7$) so that the $(p-3)$ term would match the first set of parentheses. Now factor out $(p-3)$, leaving behind $1p$ and -7.

$(p-3)(1p-7)$
$(p-3)(p-7)$

2.054 Factor the quadratic $3x^2 + 4x + 1$.

Comparing $3x^2 + 4x + 1$ with the quadratic $ax^2 + bx + c$, note that $a = 3$, $b = 4$, $c = 1$.
What's the product ac? It is $3 \times 1 = 3$.
The only way to obtain the product 3 is:

  $\left(\!\begin{array}{c}1 \times 3\end{array}\!\right) \longleftarrow$

The coefficient of middle term, 4, can be obtained by *adding* this pair.
Therefore, split the middle term as $1x + 3x$.

  $3x^2 + 1x + 3x + 1$

Group the first two terms. Group the next two terms. Factor the most you can from each group.

$$\overbrace{3x^2 + 1x} \quad \overbrace{+3x + 1}$$
$$1x(3x+1) + 1(3x+1)$$

Notice that from the second set of parentheses, I factored out 1. *Being unable to factor out anything means factoring out just 1, because 1 times any term is that term itself.*
The term $(3x+1)$ shows up in both groups. Factor it out, leaving behind $1x$ and $+1$.

  $(3x+1)(1x+1)$
  $(3x+1)(x+1)$

2.055 Factor the quadratic $5z^2 - 12z + 4$.

The variable is z, not x. Comparing $5z^2 - 12z + 4$ with the quadratic $ax^2 + bx + c$, we see that $a = 5$, $b = -12$, $c = 4$.
What's the product ac? It is $5 \times 4 = 20$.
The middle term is -12, not just 12. So, consider negative possibilities also when coming up with factors of the product 20. The different ways to obtain the product 20 are:

  $1 \times 20 \qquad (-1) \times (-20) \qquad 2 \times 10 \qquad \left(\!\begin{array}{c}(-2) \times (-10)\end{array}\!\right) \longleftarrow \quad 4 \times 5 \qquad (-4) \times (-5)$

The coefficient of middle term, -12, can be obtained by *adding* this pair.
Therefore, split the middle term as $(-2z) + (-10z)$.

  $5z^2 + (-2z) + (-10z) + 4$

Group the first two terms. Group the next two terms.

$$\overbrace{5z^2 + (-2z)} \quad \overbrace{+(-10z) + 4}$$

Simplify the excess signs between the terms. Factor the most you can from each group.

$$\overbrace{5z^2 - 2z} \quad \overbrace{-10z + 4}$$
$$1z(5z-2) - 2(5z-2)$$

From the second set of parentheses, I factored out -2 (not $+2$) so that the $(5z-2)$ term would match the first set of parentheses. Now factor out $(5z-2)$, leaving behind $1z$ and -2.

  $(5z-2)(1z-2)$
  $(5z-2)(z-2)$

2.056 Factor the quadratic $g^2 + 16g - 36$.

The variable is g, and g^2 actually means $1g^2$. Comparing $1g^2 + 16g - 36$ with the quadratic $ax^2 + bx + c$, we see that $a = 1$, $b = 16$, $c = -36$.
What's the product ac? It is $1 \times (-36) = -36$.

The different ways to obtain the product -36 are:

$1\times(-36)$ $\quad-1\times36$ $\quad2\times(-18)$ $\quad-2\times18$ $\quad3\times(-12)$ $\quad-3\times12$

$4\times(-9)$ $\quad-4\times9$ $\quad6\times(-6)$

The coefficient of middle term, 16, can be obtained by *adding* this pair.

Therefore, split the middle term as $(-2g)+18g$.

$1g^2+(-2g)+18g-36$

Group the first two terms. Group the next two terms.

$\overbrace{1g^2+(-2g)}\quad\overbrace{+18g-36}$

Simplify the excess signs between the terms. Factor the most you can from each group.

$\overbrace{1g^2-2g}\quad\overbrace{+18g-36}$

$1g(g-2)+18(g-2)$

Now factor out $(g-2)$, leaving behind $1g$ and $+18$.

$(g-2)(1g+18)$

$(g-2)(g+18)$

2.057 Factor the quadratic $4q^2+q-3$.

Note that the middle term q actually means $1q$. Comparing $4q^2+1q-3$ with the quadratic ax^2+bx+c, we see that $a=4$, $b=1$, $c=-3$.

What's the product ac? It is $4\times(-3)=-12$.

The different ways to obtain the product -12 are:

$1\times(-12)$ $\quad-1\times12$ $\quad2\times(-6)$ $\quad-2\times6$ $\quad3\times(-4)$ $\quad-3\times4$

The coefficient of middle term, 1, can be obtained by *adding* this pair.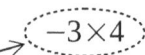

Therefore, split the middle term as $(-3q)+4q$

$4q^2+(-3q)+4q-3$

Group the first two terms. Group the next two terms.

$\overbrace{4q^2+(-3q)}\quad\overbrace{+4q-3}$

Simplify the excess signs between the terms. Factor the most you can from each group.

$\overbrace{4q^2-3q}\quad\overbrace{+4q-3}$

$1q(4q-3)+1(4q-3)$

Notice that from the second set of parentheses, I factored out 1. *Being unable to factor out anything means factoring out just 1, because 1 times any term is that term itself.*

The term $(4q-3)$ shows up in both groups. Factor it out, leaving behind $1q$ and $+1$.

$(4q-3)(1q+1)$

$(4q-3)(q+1)$

2.058 Factor the quadratic $y^2-8y-33$.

The variable is y, and y^2 actually means $1y^2$. Comparing $1y^2-8y-33$ with the quadratic ax^2+bx+c, we see that $a=1$, $b=-8$, $c=-33$.

What's the product ac? It is $1\times(-33)=-33$.

The different ways to obtain the product −33 are:

$1\times(-33)$ -1×33 $3\times(-11)$ -3×11

The coefficient of middle term, −8, can be obtained by *adding* this pair.

Therefore, split the middle term as $3y+(-11y)$.

$1y^2+3y+(-11y)-33$

Group the first two terms. Group the next two terms.

$\overbrace{1y^2+3y}$ $\overbrace{+(-11y)-33}$

Simplify the excess signs between the terms. Factor the most you can from each group.

$\overbrace{1y^2+3y}$ $\overbrace{-11y-33}$
$1y(y+3)-11(y+3)$

Now factor out $(y+3)$, leaving behind $1y$ and −11.

$(y+3)(1y-11)$
$(y+3)(y-11)$

2.059 Factor the quadratic $2m^2-13m-7$.

The variable is m. Comparing $2m^2-13m-7$ with the quadratic ax^2+bx+c, we see that $a=2,\ b=-13,\ c=-7$.

What's the product ac? It is $2\times(-7)=-14$.

The different ways to obtain the product −14 are:

$1\times(-14)$ -1×14 $2\times(-7)$ -2×7

The coefficient of middle term, −13, can be obtained by *adding* this pair.

Therefore, split the middle term as $1m+(-14m)$.

$2m^2+1m+(-14m)-7$

Group the first two terms. Group the next two terms.

$\overbrace{2m^2+1m}$ $\overbrace{+(-14m)-7}$

Simplify the excess signs between the terms. Factor the most you can from each group.

$\overbrace{2m^2+1m}$ $\overbrace{-14m-7}$
$1m(2m+1)-7(2m+1)$

The term $(2m+1)$ shows up in both groups. Factor it out, leaving behind $1m$ and −7.

$(2m+1)(1m-7)$
$(2m+1)(m-7)$

I used a different variable in each of the last eight examples to eliminate your fear of variables. Please don't expect to see the variable x every single time. Sometimes, different variables become necessary when representing different quantities. For example, m for money, p for price, r for revenue, t for time, h for height etc. I hope you get the point.

PRACTICE PROBLEMS
Factor the following polynomials.

2.060 $d^2+12d+11$ **2.062** $7h^2+12h+5$ **2.064** $r^2+48r-100$ **2.066** $u^2-23u-50$

2.061 $j^2-16j+39$ **2.063** $9k^2-15k+4$ **2.065** $12t^2+5t-2$ **2.067** $15w^2-2w-1$

FACTORING THE DIFFERENCE OF SQUARES: a^2-b^2

To understand how to factor this type of expression, let's first carry out a multiplication of a special type.

$(a+b)(a-b)$

Using the regular methodology, distribute a onto $(a-b)$. Then distribute b onto $(a-b)$.

$a\cdot(a-b)+b\cdot(a-b)$

$\overbrace{a\cdot a}\quad\overbrace{-a\cdot b}\quad\overbrace{+b\cdot a}\quad\overbrace{-b\cdot b}$

$a^2-ab+ba-b^2$

The $-ab$ and ba would cancel each other out.

We're left with a^2-b^2.

Therefore, $(a+b)(a-b)=a^2-b^2$

It can be written backwards as $a^2-b^2=(a+b)(a-b)$

The following examples show how to imitate this formula to factor *the difference of squares*.

➢ $x^2-y^2=(x+y)(x-y)$

➢ $c^2-d^2=(c+d)(c-d)$

➢ $p^2-q^2=(p+q)(p-q)$

➢ $k^2-5^2=(k+5)(k-5)$

➢ $m^2-36 = m^2-6^2 = (m+6)(m-6)$

➢ $A^2-4 = A^2-2^2 = (A+2)(A-2)$

➢ $4x^2-b^2 = 2^2x^2-b^2 = (2x+b)(2x-b)$

➢ $25n^2-81 = 5^2n^2-9^2 = (5n+9)(5n-9)$

➢ $25n^2-81m^2 = 5^2n^2-9^2m^2 = (5n+9m)(5n-9m)$

➢ $100e^2-1 = 10^2e^2-1^2 = (10e+1)(10e-1)$

I hope you get the point.

PRACTICE PROBLEMS
Factor the following using the difference of squares formula.

2.068 g^2-h^2 2.071 $16f^2-4j^2$ 2.074 $36-m^2$
2.069 L^2-R^2 2.072 a^2-49 2.075 $25x^2-1$
2.070 b^2-9c^2 2.073 $64y^2-x^2$ 2.076 $1-100k^2$

RATIONAL EXPRESSIONS

An expression in the form of a fraction with the numerator and/or the denominator being a polynomial is called a **rational expression**. Examples would be $\dfrac{x+1}{x+5}$, $\dfrac{4}{x}$, $\dfrac{x^2-3x+2}{x^2+5x-1}$, $\dfrac{x-15}{3}$ etc.

Rational expressions in algebra are handled the same way fractions are handled in arithmetic. To see how, let us study the common operations such as addition, subtraction, multiplication and division performed on rational expressions.

ADDITION/SUBTRACTION OF RATIONAL EXPRESSIONS

Back when we added or subtracted fractions (refer page 10 if need be), we first equalized their denominators (if they weren't already equal). Then we added or subtracted the numerators while leaving the denominator unchanged. We'll do the same thing here as well, except we will deal with expressions instead of pure numbers.

2.077 $\dfrac{3x+5}{x-3}+\dfrac{6-2x}{x-3}$

Notice that the denominators are identical. Add the numerators. Keep the denominator the same.

$\dfrac{3x+5+6-2x}{x-3}$

Combine the *like terms* in the numerator.

$\dfrac{1x+11}{x-3}$

$1x$ is really just x.

$\dfrac{x+11}{x+3}$

2.078 $\dfrac{x+8}{x+1}-\dfrac{6}{x+1}$

Denominators are the same. Combine the numerators. Keep the denominator unchanged.

$\dfrac{x+8-6}{x+1}$

$\dfrac{x+2}{x+1}$

2.079 $\dfrac{2x-5}{x+2}-\dfrac{x+1}{x+2}$

Denominators are the same. Combine the numerators. Keep the denominator unchanged.

$\dfrac{2x-5-(x+1)}{x+2}$

Now be sure to *distribute* the negative sign onto the terms x and 1. You may notice that this happens only when subtracting numerators.

$\dfrac{2x-5-x-1}{x+2}$

Combine the *like terms*.

$\dfrac{1x-6}{x+2}$

$\dfrac{x-6}{x+2}$

2.080 $\dfrac{x+6}{x-4}-\dfrac{3+2x}{x-4}$

Denominators are the same. Combine the numerators. Keep the denominator unchanged.

$\dfrac{x+6-(3+2x)}{x-4}$

Distribute the (−) sign onto the terms 3 and $2x$.

$\dfrac{x+6-3-2x}{x-4}$

$\dfrac{-1x+3}{x-4}$

$\dfrac{-x+3}{x-4}$

2.081 $\dfrac{2}{x}+\dfrac{1}{x-3}$

Denominators are different, their LCM being their product, $x(x{-}3)$. Please note that I've written this book to teach algebra from the ground up. The examples won't go into extraordinary level of difficulty. Thus, when it comes to finding the LCM of two algebraic expressions, in this book it will always be their product.

Multiply the top and bottom of the first rational expression by $(x{-}3)$, and the top and bottom of the second one by x. (We can't selectively multiply only the denominators by $(x{-}3)$ and x in order to equalize them; we have to multiply the numerators as well by $(x{-}3)$ and x in order to maintain the integrity of the original rational expressions.)

$$\frac{2\cdot(x-3)}{x\cdot(x-3)}+\frac{1\cdot x}{(x-3)\cdot x}$$

Now that the denominators are equal, combine the numerators over the same denominator.

$$\frac{2(x-3)+1\,x}{x(x-3)}$$

Distribute and combine the *like terms* in the numerator.

$$\frac{2\,x-6+1\,x}{x(x-3)}$$

$$\frac{3\,x-6}{x(x-3)}$$

Optionally, factor the numerator if it can be factored.

$$\frac{3(x-2)}{x(x-3)}$$

It is always a good idea to present answers in factored form when possible, because sometimes identical factors in the numerator and denominator cancel and leave the expression cleaner.

2.082 $\dfrac{3}{x-1}-\dfrac{4}{x-2}$

Denominators are different, their LCM being $(x{-}1)(x{-}2)$. Multiply the top and bottom of the first rational expression by $(x{-}2)$, and the top and bottom of the second one by $(x{-}1)$.

$$\frac{3\cdot(x-2)}{(x-1)\cdot(x-2)}-\frac{4\cdot(x-1)}{(x-2)\cdot(x-1)}$$

Combine the numerators over the same denominator.

$$\frac{3(x-2)-4(x-1)}{(x-1)(x-2)}$$

Distribute and combine *like terms* in the numerator.

$$\frac{3\,x-6-4\,x+4}{(x-1)(x-2)}$$

$$\frac{-1\,x-2}{(x-1)(x-2)}$$

$$\frac{-x-2}{(x-1)(x-2)}$$

$$\frac{-(x+2)}{(x-1)(x-2)}$$

2.083 $\dfrac{x+2}{x+8}+\dfrac{x-3}{x+1}$

Denominators are different, their LCM being $(x+8)(x+1)$. Multiply the top and bottom of the first rational expression by $(x+1)$, and the top and bottom of the second one by $(x+8)$.

$$\frac{(x+2)\cdot(x+1)}{(x+8)\cdot(x+1)}+\frac{(x-3)\cdot(x+8)}{(x+1)\cdot(x+8)}$$

Combine the numerators.

$$\frac{(x+2)(x+1)+(x-3)(x+8)}{(x+8)(x+1)}$$

Distribute and combine *like terms* in the numerator.

$$\frac{x^2+1x+2x+2+x^2+8x-3x-24}{(x+8)(x+1)}$$

$$\frac{2x^2+8x-22}{(x+8)(x+1)}$$

Factor out 2 from the numerator.

$$\frac{2(x^2+4x-11)}{(x+8)(x+1)}$$

2.084 $\dfrac{x-2}{x-5}-\dfrac{x+6}{x+4}$

Denominators are different, their LCM being $(x-5)(x+4)$. Multiply the top and bottom of the first rational expression by $(x+4)$, and the top and bottom of the second one by $(x-5)$.

$$\frac{(x-2)\cdot(x+4)}{(x-5)\cdot(x+4)}-\frac{(x+6)\cdot(x-5)}{(x+4)\cdot(x-5)}$$

Combine the numerators.

$$\frac{(x-2)\cdot(x+4)-(x+6)\cdot(x-5)}{(x+4)\cdot(x-5)}$$

Distribute and combine *like terms* in the numerator. Beware of the $(-)$ sign between the two groups of terms in the numerator. When you distribute the second set of terms, keep them in their own separate set of parentheses in this step.

$$\frac{x^2+4x-2x-8-(x^2-5x+6x-30)}{(x+4)(x-5)}$$

Now remove the parentheses in the numerator and distribute the $(-)$ sign onto the inner terms.

$$\frac{x^2+4x-2x-8-x^2+5x-6x+30}{(x+4)(x-5)}$$

$$\frac{1x+22}{(x+4)(x-5)}$$

$$\frac{x+22}{(x+4)(x-5)}$$

PRACTICE PROBLEMS

2.085 $\dfrac{8}{x}+\dfrac{3}{x}-\dfrac{2}{x}$

2.086 $\dfrac{x+7}{x-10}+\dfrac{x-2}{x-10}$

2.087 $\dfrac{2x+9}{x+7}-\dfrac{3x-4}{x+7}$

2.088 $\dfrac{x+5}{x+1}+\dfrac{x-1}{x-2}$

2.089 $\dfrac{4}{x-1}-\dfrac{x-3}{x+6}$

2.090 $\dfrac{x+4}{x-3}-\dfrac{x-2}{x+3}$

MULTIPLICATION OF RATIONAL EXPRESSIONS

This is along the lines of multiplication of fractions from arithmetic (refer page 12 if need be). When multiplying rational expressions, first factor each polynomial where possible. Then reduce and/or cross reduce any identical terms in the numerators and denominators. The examples that follow demonstrate this technique.

2.091 $\dfrac{x}{x+3} \times \dfrac{x+3}{2x-1}$

Notice that the $(x+3)$ term can be cross reduced. Please also note that it is not correct to selectively reduce the x in the numerator with the x in the denominator in either fraction. In order to reduce in rational expressions, an entire factor such as $(x+3)$ must be selected.

$$\dfrac{x}{{}_1\,\cancel{x+3}} \times \dfrac{\cancel{x+3}^{\,1}}{2x-1}$$

Multiply straight across.

$$\dfrac{x \cdot 1}{1 \cdot (2x-1)}$$

$$\dfrac{x}{2x-1}$$

2.092 $\dfrac{x^2-3x}{x+7} \times \dfrac{x+7}{5x}$

Factor the (x^2-3x) term.

$$\dfrac{x(x-3)}{x+7} \times \dfrac{x+7}{5x}$$

Cross reduce the $(x+7)$ term. Cross reduce the exclusive x from left numerator with x from right denominator. *Yes, it is correct to do so* because x is a factor of the left numerator, and x is a factor of the right denominator.

$$\dfrac{{}^1\cancel{x}(x-3)}{{}_1\,\cancel{x+7}} \times \dfrac{\cancel{x+7}^{\,1}}{5\cancel{x}_1}$$

Multiply straight across.

$$\dfrac{1 \cdot (x-3) \cdot 1}{1 \cdot 5 \cdot 1}$$

$$\dfrac{x-3}{5}$$

2.093 $\dfrac{8x^2}{4x-20} \times \dfrac{x-5}{6x}$

Factor the $(4x-20)$ term. Also, rewrite x^2 as $x \cdot x$ for easy reducing in the next step.

$$\dfrac{8x \cdot x}{4(x-5)} \times \dfrac{x-5}{6x}$$

Reduce the 8 with 4. Cross reduce $(x-5)$. Cross reduce one x from top left with x from bottom right.

$$\dfrac{{}^2\cancel{8}x \cdot \cancel{x}^{\,1}}{{}_1\cancel{4}(\cancel{x-5})_1} \times \dfrac{\cancel{x-5}^{\,1}}{6\cancel{x}^{\,1}}$$

Multiply straight across.

$$\dfrac{2 \cdot x \cdot 1 \cdot 1}{1 \cdot 1 \cdot 6 \cdot 1}$$

$$\dfrac{2x}{6}$$

Reduce 6 with 2.

$$\dfrac{x}{3}$$

2.094 $\dfrac{x^2-4}{x-2} \times \dfrac{3x+1}{x+2}$

The factorization of x^2-4 is $(x+2)(x-2)$. See page 36 on factoring the difference of squares.

$$\dfrac{(x+2)(x-2)}{x-2} \times \dfrac{3x+1}{x+2}$$

Reduce the two $(x-2)$ terms. Cross reduce the two $(x+2)$ terms.

$$\dfrac{{}^1\cancel{(x+2)}\cancel{(x-2)}^{\,1}}{\cancel{x-2}^{\,1}} \times \dfrac{3x+1}{\cancel{x+2}^{\,1}}$$

Multiply straight across.

$$\dfrac{1 \cdot 1 \cdot (3x+1)}{1 \cdot 1}$$

$$\dfrac{3x+1}{1}$$

$$3x+1$$

2.095 $\dfrac{x^2+3x+2}{x+8}\times\dfrac{2x-2}{x^2-1}$

Let's first do some individual factorizations, and then bring them all together.

➢ Factor the difference of squares, x^2-1. Refer page 36 if need be.
$$x^2-1 \;=\; x^2-1^2 \;=\; (x+1)(x-1)$$

➢ Factor the quadratic, x^2+3x+2. Refer page 31 if need be.
$$x^2+3x+2$$
$$1x^2+3x+2$$

Comparing with the quadratic ax^2+bx+c, note that $a=1$, $b=3$, $c=2$. The product $ac=2$ can only be obtained as 1×2. Split the middle term as $1x+2x$.
$$1x^2+1x+2x+2$$

Factor the first two terms. Factor the next two terms.
$$1x(x+1) \;+2(x+1)$$

The $(x+1)$ term is common to both groups. Factor it out, leaving behind $1x$ and $+2$.
$$(x+1)(1x+2)$$
$$(x+1)(x+2)$$

➢ Factor the $2x-2$ term as $2(x-1)$.

Now plug in the three factorizations into the original expression.
$$\dfrac{(x+1)(x+2)}{x+8}\times\dfrac{2(x-1)}{(x+1)(x-1)}$$

Reduce the $(x-1)$ terms. Cross reduce the $(x+1)$ terms.
$$\dfrac{^1\cancel{(x+1)}(x+2)}{x+8} \;\times\; \dfrac{2\cancel{(x-1)}^{\,1}}{^1\cancel{(x+1)}\cancel{(x-1)}^{\,1}}$$

Multiply straight across.
$$\dfrac{1\cdot(x+2)\cdot2\cdot1}{(x+8)\cdot1\cdot1}$$
$$\dfrac{2(x+2)}{(x+8)}$$

2.096 $\dfrac{x^2-49}{x^2+10x+21}\times\dfrac{2x+6}{4}$

The factorization of $2x+6$ is $2(x+3)$.

The factorization of x^2-49 is $(x+7)(x-7)$.

The factorization of $x^2+10x+21$ is:
$$x^2+10x+21$$
$$1x^2+10x+21$$
$$1x^2+3x+7x+21$$
$$1x(x+3)+7(x+3)$$
$$(x+3)(1x+7)$$
$$(x+3)(x+7)$$

Plug in these three factorizations into the original expression.
$$\dfrac{^1\cancel{(x+7)}(x-7)}{_1\cancel{(x+3)}\cancel{(x+7)}_{\,1}} \;\times\; \dfrac{^1\cancel{2}\cancel{(x+3)}^{\,1}}{\cancel{4}_{\,2}}$$
$$\dfrac{1\cdot(x-7)\cdot1\cdot1}{1\cdot1\cdot2}$$
$$\dfrac{x-7}{2}$$

PRACTICE PROBLEMS

2.097 $\dfrac{x+8}{5x}\times\dfrac{x+1}{x+8}$

2.098 $\dfrac{3x-6}{6x+6}\times\dfrac{2}{x-2}$

2.099 $\dfrac{2x-16}{x^2-64}\times\dfrac{x+8}{10x}$

2.100 $\dfrac{x^2-1}{x^2-x-2}\times\dfrac{4x-8}{12}$

2.101 $\dfrac{5}{2(x-2)}\times\dfrac{x^2-5x+6}{x^2-9}$

2.102 $\dfrac{2}{2x-18}\times\dfrac{x^2-81}{x+9}$

DIVISION OF RATIONAL EXPRESSIONS

Division of rational expressions is also along the lines of division of fractions (from page 13). Dividing one rational expression by another is the exact same as multiplying the first rational expression by the *reciprocal of the second rational expression*. Reciprocal of a rational expression, in simple language, means a rational expression turned upside down.

2.103 $\dfrac{(x+8)(x-1)}{x^2-16} \div \dfrac{x+3}{x-4}$

Multiply the first rational expression with the reciprocal of the second.
$$\dfrac{(x+8)(x-1)}{x^2-16} \times \dfrac{x-4}{x+3}$$
The factorization of x^2-16 is $(x+4)(x-4)$.
$$\dfrac{(x+8)(x-1)}{(x+4)(x-4)} \times \dfrac{x-4}{x+3}$$
Cross reduce the $(x-4)$ terms.
$$\dfrac{(x+8)(x-1)}{(x+4)\cancel{(x-4)}_1} \times \dfrac{\cancel{x-4}^1}{x+3}$$
Multiply straight across.
$$\dfrac{(x+8)(x-1)\cdot 1}{(x+4)\cdot 1 \cdot (x+3)}$$
$$\dfrac{(x+8)(x-1)}{(x+4)(x+3)}$$

2.104 $\dfrac{x+1}{5x+5} \div \dfrac{x^2-9}{10(x-3)}$

Multiply the first rational expression with the reciprocal of the second.
$$\dfrac{x+1}{5x+5} \times \dfrac{10(x-3)}{x^2-9}$$
The factorization of x^2-9 is $(x+3)(x-3)$.
The factorization of $5x+5$ is $5(x+1)$.
$$\dfrac{x+1}{5(x+1)} \times \dfrac{10(x-3)}{(x+3)(x-3)}$$
Reduce the $(x+1)$ terms. Reduce the $(x-3)$ terms. Reduce 10 with 5.
$$\dfrac{\cancel{x+1}^1}{{}_1\cancel{5(x+1)}^1} \times \dfrac{{}^2\cancel{10}(x-3)^1}{(x+3)\cancel{(x-3)}^1}$$
$$\dfrac{1\cdot 2\cdot 1}{1\cdot 1\cdot (x+3)\cdot 1}$$
$$\dfrac{2}{x+3}$$

2.105 $\dfrac{4x^2}{x^2+7x} \div \dfrac{3x}{x^2+5x-14}$

Multiply the first rational expression with the reciprocal of the second.
$$\dfrac{4x^2}{x^2+7x} \times \dfrac{x^2+5x-14}{3x}$$
Rewrite $4x^2$ as $4x\cdot x$ for easy reducing later.
The factorization of x^2+7x is $x(x+7)$.
The factorization of $x^2+5x-14$ is:
$$1x^2+5x-14$$
$$1x^2+7x-2x-14$$
$$1x(x+7)-2(x+7)$$
$$(x+7)(1x-2)$$
$$(x+7)(x-2)$$
Refer page 31 if you need a refresher on the above methodology.
The original expression now becomes:
$$\dfrac{4x\cdot x}{x(x+7)} \times \dfrac{(x+7)(x-2)}{3x}$$
$$\dfrac{4\,^1\cancel{x}\cancel{x}^1}{{}_1\cancel{x}\cancel{(x+7)}_1} \times \dfrac{{}^1\cancel{(x+7)}(x-2)}{3\cancel{x}_1}$$
$$\dfrac{4\cdot 1\cdot 1\cdot 1\cdot(x-2)}{1\cdot 1\cdot 3\cdot 1}$$
$$\dfrac{4(x-2)}{3}$$

PRACTICE PROBLEMS

2.106 $\dfrac{x-1}{x^2-2x} \div \dfrac{x^2-1}{(x-2)(x+1)}$

2.107 $\dfrac{3x+12}{x^2+9x+20} \div \dfrac{x-2}{x^2-25}$

2.108 $\dfrac{x^2-36}{x-4} \div \dfrac{2x-12}{2x-8}$

EXPONENTS

We've briefly discussed exponents earlier in the book (page 5), but now let's study them in a little more detail. This is where I feel morally obligated to disclose to you that mathematics *does* require some amount of memorizing. It's not just "practice makes perfect;" it's also "memorizing makes practicing for perfection possible." Cheesy, but true.

Let's first understand how the laws of exponents work with the help of a few simple demonstrations. We'll then apply these laws to solve slightly more complex examples.

	Law of Exponents	**How It Works**
1	$x^a \cdot x^b = x^{a+b}$	$x^4 \cdot x^5 = x^{4+5} = x^9$
2	$\dfrac{x^a}{x^b} = x^{a-b}$	$\dfrac{x^6}{x^2} = x^{6-2} = x^4$
3	$\left(x^a\right)^b = x^{a \times b}$	$\left(x^2\right)^5 = x^{2 \times 5} = x^{10}$
4	$x^{-a} = \dfrac{1}{x^a}$	$x^{-3} = \dfrac{1}{x^3}$
5	$\dfrac{1}{x^{-b}} = x^b$	$\dfrac{1}{x^{-2}} = x^2$
6	$x^0 = 1$, provided $x \neq 0$ because 0^0 is undefined.	$5^0 = 1,\ 1.64^0 = 1,\ (-8)^0 = 1,\ \left(\dfrac{3}{7}\right)^0 = 1$
7	$x^c \cdot y^c = (x \cdot y)^c$	$x^7 \cdot y^7 = (x \cdot y)^7$
8	$\dfrac{x^c}{y^c} = \left(\dfrac{x}{y}\right)^c$	$\dfrac{x^5}{y^5} = \left(\dfrac{x}{y}\right)^5$

The above laws work in both directions. Some of the examples considered below require more than one laws to simplify. Also, it is generally a good practice to write the final answer in terms of positive exponents.

Calculations should be done for numbers that aren't too big. For example 5^2 is 25; 4^3 is 64; 2^5 is 32 etc. However, most people will find 6^8 too difficult to calculate without a calculator. In such instances, leave the answer in terms of exponents instead of trying to calculate the actual value.

2.109 $x^5 \cdot x^9$

x^{5+9} Refer Law 1.

x^{14}

2.110 $\left(p^4\right)^5$

$p^{4 \times 5}$ Refer Law 3.

p^{20}

2.111 $4^7 \div 4^5$

$\dfrac{4^7}{4^5}$

4^{7-5} Refer Law 2.

4^2

4×4

16

2.112 $\left(6^3\right)^{-2}$

$6^{3\times(-2)}$ Refer Law 3.

6^{-6}

$\dfrac{1}{6^6}$ Refer Law 4.

2.113 $\left(2^4\right)^3 \div \left(2^6\right)^2$

$\dfrac{\left(2^4\right)^3}{\left(2^6\right)^2}$

$\dfrac{2^{4\times3}}{2^{6\times2}}$ Refer Law 3.

$\dfrac{2^{12}}{2^{12}}$

2^{12-12} Refer Law 2.

2^0

1 Refer Law 6.

2.114 $\dfrac{8^{-3}}{4^{-3}}$

$\left(\dfrac{8}{4}\right)^{-3}$ Refer Law 8.

2^{-3}

$\dfrac{1}{2^3}$ Refer Law 4.

$\dfrac{1}{2\times2\times2}$

$\dfrac{1}{8}$

2.115 $x^{11}\cdot x^{-6}\cdot x^4$

$x^{11+(-6)+4}$ Refer Law 1.

x^9

2.116 $a^3\times b^3 \div c^3$

$\dfrac{a^3 b^3}{c^3}$

$\dfrac{(ab)^3}{c^3}$ Refer Law 7.

$\left(\dfrac{ab}{c}\right)^3$ Refer Law 8.

2.117 $5(ky)^4$

Notice that 5 is out of the parentheses. It's not raised to 4.

$5k^4 y^4$ Refer Law 7.

2.118 $\dfrac{4(pqr)^6}{2^4}$

$\dfrac{4(pqr)^6}{2\times2\times2\times2}$

$\dfrac{4(pqr)^6}{16}$

Reduce 16 with 4.

$\dfrac{(pqr)^6}{4}$

$\dfrac{p^6 q^6 r^6}{4}$ Refer Law 7.

PRACTICE PROBLEMS

2.119 $\dfrac{x^2\cdot x^7}{p^6\cdot p^3}$

2.120 $(2m)^3 \div (4m)^2$

2.121 $p^4 \times q^4 \times r^{-4}$

2.122 $\left(7^{-2}\right)^3 \times 7^8$

2.123 $y^{-6}\times y^4$

2.124 $\left(c^8\right)^8 \div \left(c^2\right)^{32}$

2.125 $2^{10}\times2^{-6} \div 4^4$

2.126 $5^{-5}\div 5^{-2}$

CHAPTER EXERCISES

2.127 Evaluate $6x^2+\dfrac{3p}{5r}$ for
$x=-1,\ p=20,\ r=-4.$

2.128 $(10x^2-4)+(7x-3x^2)+(5-8x)$

2.129 $(-13k+14ab-9)-(2-15k+6ab)$

2.130 Distribute: $-4a(3b^2+7ab-2a^2)$

2.131 Distribute and simplify:
$(2y-9)(5-6y)$

2.132 Distribute and simplify:
$(t+11u)(3t-u)$

2.133 Factor: $8y^2-18xy$

2.134 Factor: $4pq-2qp^2+50pq^3$

2.135 Factor: $x^2-17x+30$

2.136 Factor: $2y^2-5y-12$

2.137 Factor: k^2-81

2.138 Factor: $4n^2-49m^2$

2.139 $\dfrac{4x}{x-7}+\dfrac{4-x}{x-7}$

2.140 $\dfrac{x-1}{x+3}-\dfrac{x+2}{x-8}$

2.141 $\dfrac{x^2-25}{x^2-2x-15}\times\dfrac{3x+9}{6x}$

2.142 $\dfrac{x^2-16}{5x^2+10x}\div\dfrac{x^2+2x-8}{5x^2-10x}$

2.143 $\dfrac{(a^2)^4\cdot a^{-5}}{x^{-3}\cdot x^4}$

2.144 $p^3\cdot p^{-7}\cdot(q^2)^2\cdot r^4$

CHAPTER 3
EQUATIONS

What makes an **equation** an equation is the presence of the equality sign (=). If a mathematical statement doesn't have the equality sign, then it would either be an expression (about which we learned in Chapter 2) or an inequality (about which we'll learn in Chapter 5).

Here's a sample equation: $2a=16$

Using the above sample, let's try to understand the concept of an equation by imagining two identical water bottles with the same amount of water in them, represented by each side of the equation. The amount of water in the left bottle *is equal to* the amount of water in the right bottle. If I pour 20 ml of water into each bottle, would they each still have the same amount of water? Yes.

Thus, $2a+20=16+20$ is equivalent to the above equation. Meaning, *it is permissible to add the same number to both sides of an equation.*

Instead, what if I remove 35 ml of water from both the bottles? Would they still each have the same amount of water? Yes.

Therefore, $2a-35=16-35$ is equivalent to the original equation as well. Meaning, *it is permissible to subtract the same number from both sides of an equation.*

What if I bring in 4 more identical bottles on either side, each with the same amount of water as the original bottles? I'd now have 5 bottles on either side with, as you guessed it, the same total amount of water on either side!

Thus, $2a\times5=16\times5$ is equivalent to the original equation as well. Meaning, *it is permissible to multiply both sides of an equation by the same number.*

I hope you see where this is going. Likewise, *it is permissible to divide both sides of an equation by the same number.* Finally, *it is permissible to raise both sides of an equation to the same number.*

You might reasonably conclude that *whatever you do to one side of an equation, you must do it to the other side as well.*

Equations can be of many types depending on the number of variables, the degree of the terms, the type of functions involved etc. Considering that this is not a book on advanced algebra, we will only study a few core types of equations starting the next page.

EQUATIONS IN ONE VARIABLE

An equation written in only one variable (x, y, a etc.) of the first degree is called a linear equation. Equations involving exactly two variables of the first degree are also called linear equations, but we'll study them later in this chapter.

Solving an equation for the unknown variable means completely isolating that variable onto one side of the equality sign. There should remain no traces of that variable on the other side of the equality sign. To achieve that, *use opposite operations to eliminate the unwanted terms*, one by one. Are you wondering whatever the heck that means? I thought so. Let's delve directly into examples, and I'll explain as we go.

3.001 $x+12=18$

To isolate x, the 12 must go. Note that 12 is being added on the left. The opposite of addition is subtraction. So subtract 12 from both sides of the equation to eliminate it. This is what I mean by *use opposite operations to eliminate the unwanted terms*. (You'd subtract 12 from *both sides* because whatever you do to one side of an equation, you must do it to the other side as well.)

$$x+12=18$$
$$-12\quad -12$$
$$x=6$$

3.002 $10+3x=6$

To isolate x, the 10 and the 3 must go. Which one goes first? *Independent terms that are added or subtracted always go first*. The independent 10 is positive. Meaning, it's the equivalent of 10 being added on the left. So subtract 10 from both sides of the equation. (Remember, $3x$ means 3 times x, not 3 plus x. So 3 doesn't go first.)

$$10+3x=6$$
$$-10\qquad\quad -10$$
$$3x=-4$$

$3x$ means 3 times x. To eliminate 3, divide both sides of the equation by 3 because the opposite of multiplication is division.

$$\frac{3x}{3}=\frac{-4}{3}\qquad \text{which gives us}\qquad x=-\frac{4}{3}=-1\frac{1}{3}$$

3.003 $39-y=12y$

Efficiency is the name of the game. There may be more than one ways to solve this equation, but which one involves the fewest steps? Notice that if you eliminate y from the left, you'd get both the y terms on the right which you can easily combine because they're *like terms*. The y is being subtracted on the left. So add y to both sides because the opposite of subtraction is addition.

$$39-y=12y$$
$$+y\quad +y$$
$$39=13y\qquad \text{Remember that}\quad 12y+y\quad \text{means}\quad 12y+1y. \text{ Therefore, } 13y.$$

Now divide both sides by 13 because $13y$ means 13 times y, and the opposite of multiplication is division.

$$\frac{39}{13}=\frac{13y}{13}$$
$$3=y$$

3.004 $a-14=3a-6$

The variable a appears on both sides. Let's eliminate it from the right. (You can eliminate it from the left if you want. It would still take the same number of steps because both sides have two *unlike terms*.) $3a$ on the right is positive. Meaning, it's the equivalent of $3a$ being added onto the right. So subtract it from both sides.

$$\begin{array}{l} a -14=3a-6 \\ {\scriptstyle -3a} {\scriptstyle -3a} \end{array}$$
$$-2a-14=-6$$

Now the 14 has to go. Notice that 14 is being subtracted on the left. So add it to both sides.

$$\begin{array}{l} -2a-14=-6 \\ {\scriptstyle +14} {\scriptstyle +14} \end{array}$$
$$-2a=8$$

Finally, $-2a$ means –2 times a. So divide both sides by –2.

$$\frac{-2a}{-2}=\frac{8}{-2}$$
$$a=-4$$

3.005 $5(x-4)=-3$

Distribute 5 onto inner terms.

$$5x-20=-3$$

The 20 goes first because it's an independent term being subtracted. Add 20 to both sides.

$$\begin{array}{l} 5x-20=-3 \\ {\scriptstyle +20} {\scriptstyle +20} \end{array}$$
$$5x=17$$

$5x$ means 5 times x. The opposite of multiplication is division. So divide both sides by 5.

$$\frac{5x}{5}=\frac{17}{5}$$
$$x=\frac{17}{5}=3\frac{2}{5}$$

3.006 $2(c+9)=3(c-1)$

First, carry out the distribution on both sides.

$$2c+18=3c-3$$

Eliminate $3c$ from the right by subtracting it from both sides.

$$\begin{array}{l} 2c+18=3c-3 \\ {\scriptstyle -3c} {\scriptstyle -3c} \end{array}$$
$$-c+18=-3$$

Eliminate 18 by subtracting it from both sides.

$$\begin{array}{l} -c+18=-3 \\ {\scriptstyle -18} {\scriptstyle -18} \end{array}$$
$$-c=-21$$

We're solving the equation for c, not $-c$. The quickest way to turn a negative term into positive (or positive into negative for that sake) is to multiply it by –1. Multiply both sides by –1.

$$c=21$$

3.007 $\dfrac{k}{8}+\dfrac{3k}{8}=-5$

The rational expressions on the left have the same denominator. Combine them.

$$\dfrac{k+3k}{8}=-5$$

$$\dfrac{4k}{8}=-5$$

Reduce 8 with 4.

$$\dfrac{k}{2}=-5$$

k is being divided by 2. The opposite of division is multiplication. To eliminate 2, multiply both sides by 2.

$$\dfrac{k}{2}\times 2=-5\times 2$$

$$\dfrac{2k}{2}=-10$$

$$k=-10$$

3.008 $2-\dfrac{4}{5}x=7$

Eliminate the independent 2 by subtracting it from both sides.

$$\underset{-2}{2}-\dfrac{4}{5}x=\underset{-2}{7}$$

$$-\dfrac{4}{5}x=5$$

x is being multiplied by $-\dfrac{4}{5}$. So divide both sides by $-\dfrac{4}{5}$.

$$\dfrac{-\dfrac{4}{5}x}{-\dfrac{4}{5}}=\dfrac{5}{-\dfrac{4}{5}}$$

$$x=5\times\left(-\dfrac{5}{4}\right)$$

In the last step, note that division by a fraction is the same as multiplication by its reciprocal. See *division of fractions* on page 13 if you'd need a refresher.

$$x=-\dfrac{25}{4}=-6\dfrac{1}{4}$$

3.009 $1+\dfrac{2p}{9}=8-\dfrac{4}{9}p$

Eliminate the p term from the right by adding it to both sides.

$$1+\dfrac{2p}{9}=8\underset{+\frac{4}{9}p}{}-\dfrac{4}{9}p\underset{+\frac{4}{9}p}{}$$

$$1+\dfrac{2p}{9}+\dfrac{4}{9}p=8$$

Note that $\dfrac{2p}{9}$ is the same as $\dfrac{2}{9}p$.

$$1+\dfrac{2}{9}p+\dfrac{4}{9}p=8$$

$$1+\dfrac{6}{9}p=8$$

$$1+\dfrac{2}{3}p=8$$

Subtract 1 from both sides.

$$\underset{-1}{1}+\dfrac{2}{3}p=\underset{-1}{8}$$

$$\dfrac{2}{3}p=7$$

Divide both sides by $\dfrac{2}{3}$.

$$\dfrac{\dfrac{2}{3}p}{\dfrac{2}{3}}=\dfrac{7}{\dfrac{2}{3}}$$

$$p=7\times\dfrac{3}{2}=\dfrac{21}{2}=10\dfrac{1}{2}$$

3.010 $10-\dfrac{2}{3}(x+3)=6$

Subtract 10 from both sides.

$$\underset{-10}{10}-\frac{2}{3}(x+3)=\underset{-10}{6}$$

$$-\frac{2}{3}(x+3)=-4$$

At this point, you can either distribute $-\dfrac{2}{3}$ onto the inner terms first, or you can divide both sides by $-\dfrac{2}{3}$ first. I suggest dividing by $-\dfrac{2}{3}$ first.

$$\frac{-\frac{2}{3}(x+3)}{-\frac{2}{3}}=\frac{-4}{-\frac{2}{3}}$$

$$x+3=-4\times\left(-\frac{3}{2}\right)$$

$$x+3=\frac{12}{2}$$

$$x+3=6$$

Subtract 3 from both sides.

$$\underset{-3}{x+3}=\underset{-3}{6}$$

$$x=3$$

3.011 $4+\dfrac{5}{13b}=6$

Subtract the independent 4 from both sides. Also note that $\dfrac{5}{13b}$ is *not* the same as $\dfrac{5}{13}b$.

$$\underset{-4}{4}+\frac{5}{13b}=\underset{-4}{6}$$

$$\frac{5}{13b}=2$$

$$\frac{5}{13b}=\frac{2}{1}$$

When two fractions are equal to each other, their reciprocals are also equal to each other. We will be using this concept repeatedly in this chapter.

$$\frac{13b}{5}=\frac{1}{2}$$

$$\frac{13}{5}b=\frac{1}{2}$$

Divide both sides by $\dfrac{13}{5}$.

$$\frac{\frac{13}{5}b}{\frac{13}{5}}=\frac{\frac{1}{2}}{\frac{13}{5}}$$

$$b=\frac{1}{2}\times\frac{5}{13}=\frac{5}{26}$$

3.012 $2+\dfrac{1}{2x}=-6-\dfrac{3}{2x}$

Eliminate the x term from the right by adding it to both sides.

$$2+\frac{1}{2x}=-6-\frac{3}{2x}$$
$$\underset{+\frac{3}{2x}}{}\underset{+\frac{3}{2x}}{}$$

$$2+\frac{1}{2x}+\frac{3}{2x}=-6$$

$$2+\frac{4}{2x}=-6$$

$$2+\frac{2}{x}=-6$$

Subtract 2 from both sides.

$$\underset{-2}{2}+\frac{2}{x}=\underset{-2}{-6}$$

$$\frac{2}{x}=-8$$

$$\frac{2}{x}=\frac{-8}{1}$$

Take reciprocals of both sides.

$$\frac{x}{2}=\frac{1}{-8}$$

x is being divided by 2. So multiply both sides by 2.

$$\frac{x}{2}\times2=\frac{1}{-8}\times2$$

$$\frac{2x}{2}=\frac{2}{-8}$$

$$x=\frac{2}{-8}=\frac{1}{-4}=-\frac{1}{4}$$

PRACTICE PROBLEMS

3.013 $2x-1=15$

3.014 $7y+5=26$

3.015 $-a+8=3a-12$

3.016 $\frac{2}{5}x+7=4-\frac{3}{5}x$

3.017 $15+b=5(b-8)$

3.018 $-3(m-8)=-4(3-5m)$

3.019 $2(x+5)-3=8$

3.020 $\frac{4}{7}(x+21)=6$

3.021 $3+\frac{2}{9}(c-1)=11$

3.022 $\frac{10}{3x}+4=\frac{16}{3x}$

3.023 $\frac{1}{x}+\frac{7}{x}=3-\frac{2}{x}$

PHRASES AND THEIR EQUIVALENT ALGEBRAIC EXPRESSIONS

The examples we solved in the last section involved the basic mechanics of solving equations. Let's use the skills developed thus far to solve what are called *word problems* or *story problems*. After reading them, we'd have to formulate equations and then solve them to find the unknown variable. But there's an obstacle: we don't know how to convert English phrases into algebraic expressions. To address that, here's a table that lists some commonly used phrases in story problems:

English phrase	Equivalent algebraic expression
x increased by 5 5 more than x 5 greater than x	$x+5$
x decreased by 5 5 less than x	$x-5$
The sum of a and b	$a+b$
The difference of a and b	$a-b$
The product of a and b	$a\times b$
Twice the sum of p and q	$2(p+q)$
Four times the difference of p and q	$4(p-q)$
Is Gives The result is Is the same as	$=$

3.024 A number when added to 7 gives 10. What is the number?

Your gut feeling probably tells you the answer is "so obviously 3." My gut feeling tells me I should go get lunch. Let's not go by our gut feelings, or we may end up in two different places! Since we're learning to formulate an equation that we could solve to find the unknown number, let's do it systematically. Let's not solve it all mentally.

Assume the unknown number as x (or any other variable of your choice). So, x when added to 7 would mean $x+7$. The equation would then be:

$x+7=10$

$\underset{-7\quad-7}{x+7=10}$

$x=3$

3.025 When 3 is subtracted from twice a number, the result is 11. What is the number?

How about we assume the number as y this time? Twice the number would be $2y$. When 3 is subtracted from it, it would be $2y-3$. The result is 11.

So the equation would be:

$$2y-3=11$$

$$\underset{+3 \qquad +3}{2y-3=11}$$

$$2y=14$$

$$\frac{2y}{2}=\frac{14}{2}$$

$$y=7$$

3.026 The sum of a number and 8 is 15 greater than twice the number. What is the number?

Let's assume the number as n. The sum of the number and 8 would be $n+8$. Twice the number would be $2n$. It follows that $n+8$ is 15 greater than $2n$. So if we added 15 to $2n$, then it would be equal to $n+8$.

The equation would be:

$$n+8=15+2n$$

$$\underset{-2n \qquad -2n}{n+8=15+2n}$$

$$-1n+8=15$$

$$\underset{-8 \quad -8}{-1n+8=15}$$

$$-1n=7$$

$$\frac{-1n}{-1}=\frac{7}{-1}$$

$$n=-7$$

3.027 Six less than twice a number is 14. What is the number?

Assume the number as b. Twice the number would be $2b$. Six less than that would be $2b-6$. This is given as 14.

The equation would be:

$$2b-6=14$$

$$\underset{+6 \qquad +6}{2b-6=14}$$

$$2b=20$$

$$\frac{2b}{2}=\frac{20}{2}$$

$$b=10$$

3.028 The difference of 7 and a number is the same as the sum of 17 and the number. What is the number?

Assume the number as a. The difference of 7 and the number would be $7-a$. The sum of 17 and the number would be $17+a$. These two things are apparently equal.

Write the equation as:

$$7-a=17+a$$

$$\underset{-a \qquad -a}{7-a=17+a}$$

$$7-2a=17$$

$$\underset{-7 \qquad -7}{7-2a=17}$$

$$-2a=10$$

$$\frac{-2a}{-2}=\frac{10}{-2}$$

$$a=-5$$

3.029 The product of 6 and a number is 25 greater than the number. What is the number?

Assume the number as c. The product of 6 and the number would be $6c$. This is apparently 25 greater than c. So if you add 25 to c, then that would equal $6c$.

The equation would be:

$$6c=25+c$$

$$\underset{-c \qquad -c}{6c=25+c}$$

$$5c=25$$

$$\frac{5c}{5}=\frac{25}{5}$$

$$c=5$$

PRACTICE PROBLEMS

3.030 Thirteen less than 4 times a number is 15. What is the number?

3.031 The sum of a number and 18 is 3 less than 29. What is the number?

3.032 When a number is tripled, it is 18 more than what it originally was. What was it originally?

3.033 The difference of a number and 12 is 3 more than twice the number. What is the number?

3.034 The product of a number and 5 is the same as the sum of the number and 5. What is the number?

RATIO AND PROPORTION AS EQUATIONS

Ratios can be used to compare two or more quantities. If I said that I'm ¾ as tall as you are, I'd be comparing our heights. The fraction ¾ can be written as the ratio 3:4. Meaning, the ratio of my height to your height is 3:4. Likewise, if my dog is ¼ as tall as you are, then the ratio of my dog's height to your height is 1:4. We can write the combined ratio of my dog's height to my height to your height as 1:3:4. In this book, we will limit our discussion to the ratio of two quantities only, not three or more quantities.

A **proportion** is when two ratios are expressed equal to each other. For example, $4:5=8:10$ is a proportion which can be written as the equation $\frac{4}{5}=\frac{8}{10}$. Reducing $\frac{8}{10}$ gets us $\frac{4}{5}$. That's how we know they're equal.

What if we are to solve the proportion $2:3=10:x$ to find the value of x? We'd first write it as the equation $\frac{2}{3}=\frac{10}{x}$. In the last section, we learned how to solve such equations: the x is in the denominator, so we'd take reciprocals of both sides to bring it to the numerator.

$$\frac{3}{2}=\frac{x}{10}$$

Multiply both sides by 10.

$$\frac{3}{2}\times 10=\frac{x}{10}\times 10$$

$$\frac{30}{2}=\frac{10x}{10}$$

$$15=x$$

3.035 $\frac{7}{5}=\frac{28}{x}$

x is in the denominator. Take reciprocals of both sides.

$$\frac{5}{7}=\frac{x}{28}$$

Multiply both sides by 28.

$$\frac{5}{7}\times 28=\frac{x}{28}\times 28$$

$$\frac{5\times 28}{7}=\frac{28x}{28}$$

$$\frac{5\times 28\,^4}{7\,^1}=x$$

$$\frac{5\times 4}{1}=x$$

$$20=x$$

3.036 $\dfrac{2}{3x} = \dfrac{1}{6}$

$3x$ is in the denominator. Take reciprocals of both sides to bring it up.

$$\dfrac{3x}{2} = \dfrac{6}{1}$$

Notice that $\dfrac{3x}{2}$ is the same as $\dfrac{3}{2}x$.

and $\dfrac{6}{1}$ is just 6.

$$\dfrac{3}{2}x = 6$$

Divide both sides by $\dfrac{3}{2}$.

$$\dfrac{\dfrac{3}{2}x}{\dfrac{3}{2}} = \dfrac{6}{\dfrac{3}{2}}$$

$$x = 6 \times \dfrac{2}{3} = \dfrac{12}{3} = 4$$

3.037 $\dfrac{8}{6} = \dfrac{2a}{5}$

Notice that $\dfrac{2a}{5}$ is the same as $\dfrac{2}{5}a$.

$$\dfrac{8}{6} = \dfrac{2}{5}a$$

Divide both sides by $\dfrac{2}{5}$.

$$\dfrac{\dfrac{8}{6}}{\dfrac{2}{5}} = \dfrac{\dfrac{2}{5}a}{\dfrac{2}{5}}$$

$$\dfrac{\overset{4}{8}}{6} \times \dfrac{5}{\underset{1}{2}} = a$$

$$\dfrac{20}{6} = a$$

$$a = \dfrac{10}{3} = 3\dfrac{1}{3}$$

3.038 The ratio of Paul's age to Fiona's age is 6:5. If Paul is 30 years old, then how old is Fiona?

Assume Paul's age as p, and Fiona's age as f. Then, $p:f = 6:5$. But it's given that $p = 30$. The proportion would then be $30:f = 6:5$.

The equation would be:

$$\dfrac{30}{f} = \dfrac{6}{5}$$

f is in the denominator. Take reciprocals of both sides.

$$\dfrac{f}{30} = \dfrac{5}{6}$$

Multiply both sides by 30.

$$\dfrac{f}{30} \times 30 = \dfrac{5}{6} \times 30$$

$$\dfrac{30f}{30} = \dfrac{5 \times \overset{5}{30}}{\underset{1}{6}}$$

$$f = \dfrac{5 \times 5}{1} = 25 \text{ years}$$

3.039 My Reading and Math scores on a test were in the ratio of 4:5. If my Math score was 800, then what was my Reading score?

Assume R as my reading score and M as my math score. (I'm using capital letters just for a change. It doesn't necessarily have to be that way.) Then, $R:M = 4:5$. It is given that $M = 800$. The proportion would then be $R:800 = 4:5$.

The equation would be:

$$\dfrac{R}{800} = \dfrac{4}{5}$$

Multiply both sides by 800.

$$\dfrac{R}{800} \times 800 = \dfrac{4}{5} \times 800$$

$$\dfrac{800R}{800} = \dfrac{4 \times \overset{160}{800}}{\underset{1}{5}}$$

$$R = \dfrac{4 \times 160}{1} = 640$$

3.040 The unit prices of gasoline and diesel are in the ratio 9:10. If gasoline costs $2.70, what is the price of diesel?

Assume G as the price of gasoline and D as the price of diesel. Then, $G:D=9:10$. It is given that $G=2.70$. The proportion would then be $2.70:D=9:10$.

The equation would be:

$$\frac{2.70}{D}=\frac{9}{10}$$

Take reciprocals of both sides.

$$\frac{D}{2.70}=\frac{10}{9}$$

Multiply both sides by 2.70.

$$\frac{D}{2.70}\times 2.70=\frac{10}{9}\times 2.70$$

$$\frac{2.70\,D}{2.70}=\frac{10\times 2.70}{9}$$

$$D=\frac{27}{9}=3=\$3.00$$

PRACTICE PROBLEMS

3.041 $\dfrac{m}{3}=\dfrac{7}{5}$

3.042 $\dfrac{10}{3}=\dfrac{5d}{6}$

3.043 $\dfrac{4}{9y}=\dfrac{3}{2}$

3.044 The ratio of height to width of a door is $9:2$. If the height is 72 inches, then what is the width?

3.045 The ratio of number of men to number of women present at a meeting is $3:5$. If there are 12 men present at the meeting, then how many women are there?

3.046 Gary's English and History scores are in the ratio $8:7$. If his English score is 96, then what is his History score?

EQUATIONS IN TWO OR MORE VARIABLES (LITERAL EQUATIONS)

A **literal equation** contains two or more *letters*; thus the name. Letters, after all, represent variables. All formulas are literal equations, the applications of which can be found in just about any subject. For example, $E=mC^2$ in quantum physics relates the conversion between mass (m) and energy (E), where C is the velocity of light in empty space. Another common example in science is $F=1.8C+32$ where F is the temperature in degrees Fahrenheit and C is the temperature in degrees Celcius.

Any literal equation can be solved for any of the variables in it. Recall from page 47 that *solving for a variable* means completely isolating the variable onto one side of the equality sign. There shouldn't remain any traces of that variable onto the other side.

Consider the equation $2x+8y=a+by$. Let's solve it for b. Notice that the b term is only on the right, which means we should eliminate a and y from the right. a goes first; y goes next. Subtract a from both sides.

$$2x+8y \underset{-a \quad -a}{\quad=a+by}$$

$$2x+8y-a=by$$

There are no *like terms* to combine on the left. Divide both sides by y.

$$\frac{2x+8y-a}{y}=\frac{by}{y}$$

$$\frac{2x+8y-a}{y}=b \quad \text{The equation is now fully solved for } b.$$

What if we're asked to solve the original equation, $2x+8y=a+by$, for x instead? Then, the terms 2 and $8y$ would have to go from the left. $8y$ would go first; 2 would go next. Subtract $8y$ from both sides.

$$2x+8y=a+by$$
$${-8y}{-8y}$$
$$2x=a+by-8y$$

Divide both sides by 2.

$$\frac{2x}{2}=\frac{a+by-8y}{2}$$

$$x=\frac{a+by-8y}{2}\qquad\text{The equation is now fully solved for } x.$$

You can optionally factor out the y.

$$x=\frac{a+y(b-8)}{2}$$

What if the original equation, $2x+8y=a+by$, were to be solved for y? Notice that the y term appears on both sides. Eliminate it from the right by subtracting by from both sides.

$$2x+8y=a+by$$
$${-by}{-by}$$
$$2x+8y-by=a$$

Now that the y terms are on only on the left, eliminate $2x$ from the left by subtracting it from both sides.

$$2x+8y-by=a$$
$${-2x}{-2x}$$
$$8y-by=a-2x$$

To eliminate 8 and b from the left, factoring out the y is an absolute necessity; it's not just an option.

$$y(8-b)=a-2x$$

Divide both sides by $(8-b)$.

$$\frac{y(8-b)}{8-b}=\frac{a-2x}{8-b}$$

$$y=\frac{a-2x}{8-b}\qquad\text{The equation is now fully solved for } y.$$

3.047 Solve the formula $8R=a+b$ for R.

To solve it for R, the 8 must go. Divide both sides by 8.

$$\frac{8R}{8}=\frac{a+b}{8}$$

$$R=\frac{a+b}{8}$$

3.048 Solve the formula $8\pi R=a+b$ for R.

Slightly different example than the one above, but the basic methodology is the same. Please note that $8\pi R$ means $8\times\pi\times R$. Meaning, to eliminate 8π, divide both sides by it.

$$\frac{8\pi R}{8\pi}=\frac{a+b}{8\pi}$$

$$R=\frac{a+b}{8\pi}$$

3.049 Solve the equation $-7x+3y=5b$ first for x. Then start over and solve for y.

To solve for x, the -7 and $3y$ must go from the left. $3y$ goes first; -7 goes next. Subtract $3y$ from both sides.

$$-7x+3y=5b$$
$$\underset{-3y}{}\underset{-3y}{}$$
$$-7x=5b-3y$$

Divide both sides by -7.

$$\frac{-7x}{-7}=\frac{5b-3y}{-7}$$
$$x=\frac{5b-3y}{-7}$$

To solve the original equation for y, the 3 and $-7x$ must go from the left. $-7x$ goes first; 3 goes next. Add $7x$ to both sides.

$$-7x+3y=5b$$
$$\underset{+7x}{}\underset{+7x}{}$$
$$3y=5b+7x$$

Divide both sides by 3.

$$\frac{3y}{3}=\frac{5b+7x}{3}$$
$$y=\frac{5b+7x}{3}$$

3.050 Solve the equation $ax+by=3x+4$ for x.

The x terms occur on both sides. Eliminate the one on the right by subtracting it from both sides.

$$ax+by=3x+4$$
$$\underset{-3x}{}\underset{-3x}{}$$
$$ax-3x+by=4$$

Eliminate by from the left by subtracting it from both sides.

$$ax-3x+by=4$$
$$\underset{-by}{}\underset{-by}{}$$
$$ax-3x=4-by$$

Factor out x on the left.

$$x(a-3)=4-by$$

Divide both sides by $(a-3)$.

$$\frac{x(a-3)}{a-3}=\frac{4-by}{a-3}$$
$$x=\frac{4-by}{a-3}$$

3.051 Solve the formula $\frac{a}{c}+3=k+1$ for c.

To solve for c, the 3 and a must go from the left. 3 goes first; a goes next. Subtract 3 from both sides.

$$\frac{a}{c}+3=k+1$$
$$\phantom{\frac{a}{c}}\underset{-3}{}\underset{-3}{}$$
$$\frac{a}{c}=k-2$$
$$\frac{a}{c}=\frac{k-2}{1}$$

c is in the denominator. Take reciprocals of both sides to bring it to the numerator.

$$\frac{c}{a}=\frac{1}{k-2}$$

Multiply both sides by a.

$$\frac{c}{a}\times a=\frac{1}{k-2}\times a$$
$$\frac{ca}{a}=\frac{a}{k-2}$$
$$c=\frac{a}{k-2}$$

PRACTICE PROBLEMS

3.052 Solve $-7x+3y=5b$ for b.

3.053 Solve $v=u+at$ for t.

3.054 Solve $2(x+y)=18$ for x.

3.055 Solve $n-ab=3ab+m$ for a.

3.056 Solve $3p+5q=mp-q$ for p.

3.057 Solve $\frac{F}{m}=a+g$ for m.

LINEAR EQUATIONS

Earlier in this chapter, we studied linear equations in one variable. Now let's discuss those in two variables. Specifically, we are going to consider a system of two linear equations, each involving the same two variables (for example, x and y). We will solve the system to find the values of those variables. There are two distinct methods we are going to study: method of substitution, and method of elimination.

METHOD OF SUBSTITUTION

This method is preferred when at least one of the equations has *one variable expressed in the form of the other*. The clearest way to understand this method is through examples.

3.058 Solve the system of equations: $y=2x-1$ and $3x-y=4$

Notice that in the first equation, y is expressed as *other things* (being $2x-1$). It's those other things that can be substituted for y in the second equation. Therefore, it's called the **method of substitution**.

Plug in $2x-1$ in the place of y in the second equation.
$$3x-y=4$$
$$3x-(2x-1)=4$$
Distribute the outer $(-)$ sign onto the inner terms.
$$3x-2x+1=4$$
$$1x+1=4$$
$$\begin{array}{r} 1x+1=4 \\ {\scriptstyle -1 \quad -1} \end{array}$$
$$1x=3$$
$$x=3$$

Now that x is known, plug its value into the first equation to find the value of y. (Please note that you can plug it into the second equation if you'd like. You'll get the same answer, but will need more steps. Efficiency is the name of the game, remember?)
$$y=2x-1$$
$$y=2\cdot(3)-1$$
$$y=5$$

3.059 Solve the system of equations: $2x-8y=14$ and $x=2y+5$

In the second equation, x is expressed in the form of other terms (being $2y+5$). Substitute those terms for x in the first equation.
$$2x-8y=14$$
$$2(2y+5)-8y=14$$
$$4y+10-8y=14$$
$$-4y+10=14$$
$$\begin{array}{r} -4y+10=14 \\ {\scriptstyle -10 \quad -10} \end{array}$$

$$-4y=4$$

$$\frac{-4y}{-4}=\frac{4}{-4}$$

$$y=-1$$

Now that y is known, plug it in the second equation to find the value of x.

$$x=2y+5$$

$$x=2(-1)+5$$

$$x=3$$

3.060 Solve the system of equations: $x+y=-6$ and $y=x+4$

In the second equation, y is expressed in the form of other terms (being $x+4$). Substitute those terms for y in the first equation.

$$x+y=-6$$

$$x+(x+4)=-6$$

$$2x+4=-6$$

$$2x+4=-6$$
$$\quad -4 \quad -4$$

$$2x=-10$$

$$\frac{2x}{2}=\frac{-10}{2}$$

$$x=-5$$

Now that x is known, plug it in the second equation to find the value of y.

$$y=x+4$$

$$y=-5+4$$

$$y=-1$$

3.061 Martha's age is two years more than Stewart's age. If the sum of their ages is 26, then what are their ages?

It's really about being able to form the two equations. After that, just solve them like in the previous problems. Assume m as Martha's age, and s as Stewart's age. From the first sentence of the problem, we can write the first equation as $m=s+2$. From the second sentence, we can write the second equation as $m+s=26$.

In the first equation, m is expressed in the form of other terms (being $s+2$). Substitute those terms for m in the second equation.

$$m+s=26$$

$$(s+2)+s=26$$

$$2s+2=26$$

$$2s+2=26$$
$$\quad -2 \quad -2$$

$$2s=24$$

$$\frac{2s}{2}=\frac{24}{2}$$

$$s=12$$

Now that s is known, plug it in the first equation to find the value of m.

$m+s=26$

$m+12=26$

$m+12=26$
${-12}{-12}$

$m=14$

Martha's age is 14 years, and Stewart's age is 12 years.

3.062 The price of a smart phone is $30 more than the price of a basic phone. If 4 basic phones and 2 smart phones together cost $150, then what is the price of each smart phone and basic phone?

Assume S as the price of a smart phone, and B as the price of a basic phone. From the first sentence of the problem, $S=B+30$. From the second sentence, 4 basic phones would cost $4B$ dollars whereas 2 smart phones would cost $2S$ dollars. That total cost is given as $150. So the second equation would be $4B+2S=150$.

In the first equation, S is expressed in the form of other terms (being $B+30$). Substitute those terms for S in the second equation.

$4B+2S=150$

$4B+2(B+30)=150$

$4B+2B+60=150$

$6B+60=150$

$6B+60=150$
${-60}{-60}$

$6B=90$

$\dfrac{6B}{6}=\dfrac{90}{6}$

$B=15$

Now that B is known, plug it in the first equation to find the value of S.

$S=B+30$

$S=15+30$

$S=45$

A smart phone costs $45, and a basic phone costs $15.

PRACTICE PROBLEMS

3.063 Solve the system of equations: $x+y=0$ and $x=7-2y$

3.064 Solve the system: $y=9+3x$ and $2x-y=-8$

3.065 Solve the equations: $-4y+2x=20$ and $x=5y+1$

3.066 Solve: $y=-x-7$ and $5x=2y$

3.067 A notebook costs $5 less than the cost of 3 pens. Six notebooks and six pens together cost $18. How much does each pen and notebook cost?

3.068 A glass bowl is as heavy as 3 plastic bowls. If 5 plastic bowls and 2 glass bowls together weigh 66 ounces, how much does each plastic bowl and glass bowl weigh?

METHOD OF ELIMINATION

Each of the examples in the previous section had a linear equation in which one variable was expressed in the form of other terms. But what if you're given two linear equations, neither of which has such a layout? Then you'd multiply either one or both equations with appropriate constants in such a way that either the x terms or the y terms would become opposites of each other, for example $2x$ in one equation versus $-2x$ in the other, or $5y$ in one equation versus $-5y$. You would then add the equations to *eliminate* that variable (either x or y). Therefore, it's called the **method of elimination**.

3.069 Solve the system: $2x+5y=-7$ and $2x-y=11$.

Just the way we learned that it's permissible to add the same number to both sides of an equation (from page 46), it's also permissible to add the same term (such as $8x$ or $-500y$ or $29z$ or 16 etc.) to both sides of an equation. By the same logic, it's permissible to add the two sides of one equation to their corresponding sides of another equation. Meaning, add left side to left side; add right side to right side. The point of this discussion will be clear in just a minute.

$2x$ is common to both equations above. So let's try to eliminate it by converting one of them into a negative and then adding the two equations. To do that, multiply both sides of the second equation by -1. (If you'd like, you can multiply the first equation by -1 and proceed. In the end, you'll get the same answers.)

$-1\times(2x-y)=-1\times11$

$\begin{aligned}-2x+\ y&=-11\\+\ \underline{2x+5y=-7}\\6y&=-18\end{aligned}$ Bring down the first equation and add the two equations.

$\dfrac{6y}{6}=\dfrac{-18}{6}$

$y=-3$

Now that y is known, plug it into the second equation (because it looks easier to me) to find the value of x.

$2x-y=11$

$2x-(-3)=11$

$2x+3=11$

$2x+3=11$
$\quad\ {\scriptstyle-3}\quad\ {\scriptstyle-3}$

$2x=8$

$\dfrac{2x}{2}=\dfrac{8}{2}$

$x=4$

3.070 Solve the system: $x+3y=22$ and $5x+6y=38$.

Let's first decide whether to eliminate the x terms or the y terms. It's easier to eliminate the y terms by multiplying the first equation by -2 so that $3y$ turns into $-6y$ which is the opposite of $6y$ from the second equation. If we were to eliminate the x terms instead, we'd have to multiply the first equation by -5. Let's keep the numbers small to avoid using a calculator.

Multiply both sides of the first equation by -2.

$-2\times(x+3y)=-2\times22$

$$-2x-6y=-44$$ Bring down the second equation and add the two equations.
$$+\ \underline{5x+6y=\ 38}$$
$$3x\qquad=-6$$

$$\frac{3x}{3}=\frac{-6}{3}$$

$$x=-2$$

Now that x is known, plug it into the first equation to find the value of y.

$$x+3y=22$$
$$-2+3y=22$$
$$\underset{+2\qquad\ \ +2}{-2+3y=22}$$
$$3y=24$$
$$\frac{3y}{3}=\frac{24}{3}$$
$$y=8$$

3.071 Solve the system: $5x+4y=-7$ and $4x+3y=-6$.

Eliminating either the x terms or the y terms will take about the same effort. Let's target the x terms, their LCM being $20x$. Multiply both sides of the first equation by –4 (to get –$20x$), and both sides of the second equation by 5 (to get $20x$).

$$-4\times(5x+4y)=-4\times(-7)\qquad\Big|\qquad 5\times(4x+3y)=5\times(-6)$$
$$-20x-16y=28\qquad\qquad\Big|\qquad 20x+15y=-30$$

Now add the two equations.

$$-20x-16y=\ 28$$
$$+\ \underline{20x+15y=-30}$$
$$-y=-2$$

Multiply both sides by –1 to reverse the signs.

$$y=2$$

Now that y is known, plug it into the first equation to find the value of x.

$$5x+4y=-7$$
$$5x+4(2)=-7$$
$$5x+8=-7$$
$$\underset{-8\qquad\ -8}{5x+8=-7}$$
$$5x=-15$$
$$\frac{5x}{5}=\frac{-15}{5}$$
$$x=-3$$

3.072 Five sandwiches and 3 hamburgers together cost \$21, whereas 4 sandwiches and 4 hamburgers cost \$20. What is the price of each sandwich and hamburger?

Assume S as the price of a sandwich, and H as the price of a hamburger. The first part of the sentence means $5S+3H=21$. The second part means $4S+4H=20$. Let's eliminate the H terms, their LCM being $12H$.

Multiply both sides of the first equation by –4 (to get –12H), and both sides of the second equation by 3 (to get 12H).

$$-4\times(5S+3H)=-4\times21 \qquad 3\times(4S+4H)=3\times20$$
$$-20S-12H=-84 \qquad 12S+12H=60$$

Now add the two equations.

$$\begin{array}{r} -20S-12H=-84 \\ +\ 12S+12H=\ \ 60 \\ \hline -8S \qquad\ \ =-24 \end{array}$$

$$\frac{-8S}{-8}=\frac{-24}{-8}$$

$$S=3$$

Now that S is known, plug it into the first equation to find the value of H.

$$5S+3H=21$$
$$5(3)+3H=21$$
$$15+3H=21$$
$$\underset{-15\qquad\quad -15}{15+3H=21}$$
$$3H=6$$
$$\frac{3H}{3}=\frac{6}{3}$$
$$H=2$$

A sandwich costs \$3, and a hamburger costs \$2.

3.073 Three loaves of bread and a bottle of water together cost \$10. Two loaves of bread and 3 bottles of water cost \$9. What is the price of each loaf of bread and bottle of water?

Assume L as the price of a loaf of bread, and B as the price of a bottle of water. The first sentence of the problem means $3L+B=10$. The second sentence means $2L+3B=9$. Let's eliminate the B terms, their LCM being $3B$. Multiply the first equation by –3.

$$-3\times(3L+B)=-3\times10$$

$$\begin{array}{r} -9L-3B=-30 \\ +\ 2L+3B=\ \ 9 \\ \hline -7L \qquad\ =-21 \end{array}$$ Bring down the second equation and add the two equations.

$$\frac{-7L}{-7}=\frac{-21}{-7}$$

$$L=3$$

Now that L is known, plug it into the first equation to find the value of B.

$$3L+B=10$$
$$3(3)+B=10$$
$$9+B=10$$
$$\underset{-9\qquad\quad -9}{9+B=10}$$
$$B=1$$

A loaf of bread costs \$3, and a bottle of water costs \$1.

PRACTICE PROBLEMS

3.074 Solve the system: $5x-3y=13$ and $x-3y=-7$

3.075 Solve the system: $3x+2y=8$ and $5x-7y=-28$

3.076 Solve the system: $3x-7y=14$ and $x-4y=8$

3.077 A T-shirt and a sweater together cost \$21. Four T-shirts and two sweaters cost \$54. How much does each T-shirt and sweater cost?

3.078 The combined length of 3 pencils and 6 erasers is 30 inches, whereas that of 5 pencils and 4 erasers is 38 inches. What is the length of each pencil and eraser?

ABSOLUTE VALUE EQUATIONS

Absolute value of a number, also referred to as **modulus** of a number, is the value of the number without regard to its sign. Absolute values are always positive or zero, but never negative. If a number has a (–) sign in front of it, then to find its modulus, drop the sign and pick the number. For example, when finding the absolute value of –6, written as $|-6|$, drop the (–) sign, which makes it 6. Mathematically, $|-6|=6$. Similarly, $|28|=28$, $|-3.14|=3.14$, $|7|=7$, $|0|=0$ etc.

When dealing with variables, we use the same idea. But before we try to find the absolute value of a variable, how do we know if the variable has a positive or negative value? For example, consider the modulus of x, or $|x|$. If $x=5$, then $|x|=5$ as well. But if $x=-12$, then $|x|$ has to be positive regardless of the negative sign. To get that positive value, we would take the negative of –12. Meaning, $|x|=-(-12)=12$.

We can therefore define absolute value as:

$|x|=x$ if x is already positive   For example, $x=5$ means $|x|=|5|=5$

$|x|=-(x)$ if x is negative   For example, $x=-12$ means $|x|=|-12|=-(-12)=12$

An **absolute value equation** is one that involves a modulus and other terms, for example, $|x-9|=2$. Since we don't know whether x has a positive or negative value, we have to consider both the possibilities from the above definition: what's inside the modulus, and negative of what's inside the modulus. So write *two separate equations* and solve them to find the value or values of x.

$$x-9=2$$
$$\begin{array}{cc} x-9=2 \\ {\scriptstyle +9 \quad +9} \\ x=11 \end{array}$$

$$-(x-9)=2$$
$$-x+9=2$$
$$\begin{array}{cc} -x+9=2 \\ {\scriptstyle -9 \quad -9} \\ -x=-7 \\ x=7 \end{array}$$

3.079 Solve $|x+7|=8$.

$$x+7=8$$
$$\begin{array}{cc} x+7=8 \\ {\scriptstyle -7 \quad -7} \\ x=1 \end{array}$$

$$-(x+7)=8$$
$$-x-7=8$$
$$\begin{array}{cc} -x-7=8 \\ {\scriptstyle +7 \quad +7} \\ -x=15 \\ x=-15 \end{array}$$

3.080 Solve $|2x-1|=15$.

$$2x-1=15$$
$$2x-1=15$$
$$\quad\;_{+1}\quad_{+1}$$
$$2x=16$$
$$\frac{2x}{2}=\frac{16}{2}$$
$$x=8$$

$$-(2x-1)=15$$
$$-2x+1=15$$
$$-2x+1=15$$
$$\quad\;_{-1}\quad_{-1}$$
$$-2x=14$$
$$\frac{-2x}{-2}=\frac{14}{-2}$$
$$x=-7$$

3.081 Solve $3+2|x+5|=7$.

First, get the absolute value (the modulus) by itself.

$$3+2|x+5|=7$$
$$_{-3}\qquad\quad_{-3}$$
$$2|x+5|=4$$
$$\frac{2|x+5|}{2}=\frac{4}{2}$$
$$|x+5|=2$$

Now solve by writing two separate equations.

$$x+5=2$$
$$x+5=2$$
$$\;_{-5}\quad_{-5}$$
$$x=-3$$

$$-(x+5)=2$$
$$-x-5=2$$
$$-x-5=2$$
$$\quad\;_{+5}\quad_{+5}$$
$$-x=7$$
$$x=-7$$

3.082 Solve $5|27-x|+23=8$.

First, get the absolute value by itself.

$$5|27-x|+23=8$$
$$_{-23}\quad\;_{-23}$$
$$5|27-x|=-15$$
$$\frac{5|27-x|}{5}=\frac{-15}{5}$$
$$|27-x|=-3$$

We have a problem: The absolute value in this equation is negative (–3). Absolute value can be positive or zero, but *never* negative. Meaning, this equation is invalid and has no solution. This is why it is very important to first get the modulus by itself.

PRACTICE PROBLEMS
Solve the following absolute value equations:

3.083 $|14+x|=5$

3.084 $|45-3x|=33$

3.085 $4|5x-1|-1=19$

3.086 $|x+6|+12=7$

3.087 $-13+|x-15|=9$

3.088 $11+2|x+10|=5$

QUADRATIC EQUATIONS

This section builds on top of the discussion we had on factoring quadratics in the previous chapter (refer page 31). A **quadratic equation** results when a quadratic of the form ax^2+bx+c is equated to *zero*. Recall that x is the variable, whereas a, b and c are constants with known values. To solve a quadratic equation of the standard form $ax^2+bx+c=0$, first factor the quadratic (if it can be factored). Then equate each factor to zero, and solve for x. You may get up to two answers.

If the quadratic cannot be factored, then use the **quadratic formula**, given as:

$$x=\frac{-b\pm\sqrt{b^2-4ac}}{2a}$$

The $\pm$ symbol means perform calculations first with (+), then with (−). That's how you'd get the two answers.

3.089 Solve the quadratic equation $x^2+15x+56=0$.

At this point, I'm assuming you've learned how to factor a quadratic (see page 31 if needed).

$1x^2+15x+56=0$

$1x^2+7x+8x+56=0$

$1x(x+7)+8(x+7)=0$

$(x+7)(1x+8)=0$

$(x+7)(x+8)=0$

Let's step aside for a minute and try to understand the concept of what's called a **zero product**: What if I told you I'm thinking of a number which when multiplied by 25 gives zero? You'd tell me that that number would be zero itself, right? You'd be correct!

Now what if I told you I'm thinking of two numbers which when multiplied by each other give zero? You'd tell me that either the first number is zero, or the second number is zero, or *both* numbers are zero. Again, you'd be correct!

Now take a fresh look at $(x+7)(x+8)=0$. We see that $(x+7)$ is being multiplied by $(x+8)$ to give zero. Unfortunately, we don't know the value of x; so we can't tell which one of those terms amounts to zero. The most we can do is guess that either of them could *possibly* be zero. Meaning, we can set them equal to zero and solve for the values of x.

$x+7=0$	$x+8=0$
$x+7=0$	$x+8=0$
$\quad{}_{-7}\quad{}_{-7}$	$\quad{}_{-8}\quad{}_{-8}$
$x=-7$	$x=-8$

3.090 Solve the equation $4x^2+9x+5=0$.

$4x^2+4x+5x+5=0$

$4x(x+1)+5(x+1)=0$

$(x+1)(4x+5)=0$

This is a zero product. Equate each factor to zero, and solve.

	$4x+5=0$
	$\quad{}_{-5}\quad{}_{-5}$
$x+1=0$	$4x=-5$
$\quad{}_{-1}\quad{}_{-1}$	$\dfrac{4x}{4}=\dfrac{-5}{4}$
$x=-1$	$x=\dfrac{-5}{4}=-\dfrac{5}{4}=-1\dfrac{1}{4}$

3.091 Solve $x^2-7x-16=2$.

Notice that the quadratic is equal to 2, not 0. That won't work. First, eliminate 2 from the right.

$$x^2-7x-16 \underset{-2}{=} \underset{-2}{2}$$

$$x^2-7x-18=0$$

Now that the quadratic is equal to 0, factor and solve the regular way.

$$1x^2-7x-18=0$$
$$1x^2-9x+2x-18=0$$
$$1x(x-9)+2(x-9)=0$$
$$(x-9)(1x+2)=0$$
$$(x-9)(x+2)=0$$

This is a zero product. Equate each factor to zero, and solve.

$$x-9 \underset{+9}{=} \underset{+9}{0} \qquad\qquad x+2 \underset{-2}{=} \underset{-2}{0}$$

$$x=9 \qquad\qquad\qquad\qquad x=-2$$

3.092 Solve $3x^2-5x+8=17-x^2$.

Eliminate the terms from the right so that the quadratic is equal to zero.

$$3x^2-5x+8 \underset{+x^2}{=} 17 \underset{+x^2}{-x^2}$$

$$4x^2-5x+8 \underset{-17}{=} \underset{-17}{17}$$

$$4x^2-5x-9=0$$
$$4x^2+4x-9x-9=0$$
$$4x(x+1)-9(x+1)=0$$
$$(x+1)(4x-9)=0$$

This is a zero product. Equate each factor to zero, and solve.

$$x+1 \underset{-1}{=} \underset{-1}{0} \qquad\qquad 4x-9 \underset{+9}{=} \underset{+9}{0}$$

$$x=-1 \qquad\qquad\qquad 4x=9$$

$$\qquad\qquad\qquad\qquad \frac{4x}{4}=\frac{9}{4}$$

$$\qquad\qquad\qquad\qquad x=\frac{9}{4}=2\frac{1}{4}$$

3.093 Solve $x^2+7x+8=0$.

The quadratic in the equation can't be factored. So use the quadratic formula instead. (You can use the formula even if the quadratic *can* be factored. You'd get the same answers.)

Comparing $x^2+7x+8=0$ with $ax^2+bx+c=0$, we see that $a=1$, $b=7$, $c=8$.

$$x=\frac{-b\pm\sqrt{b^2-4ac}}{2a}$$

$$x=\frac{-7\pm\sqrt{7^2-4(1)(8)}}{2(1)} \qquad\qquad \text{Don't forget PEMDAS, the order of operations (refer page 6).}$$

$$x = \frac{-7 \pm \sqrt{49-32}}{2}$$

$$x = \frac{-7 \pm \sqrt{17}}{2}$$

The two answers are: $x = \frac{-7+\sqrt{17}}{2}$ and $x = \frac{-7-\sqrt{17}}{2}$

Rounded to two decimal digits, the answers are: $x = -1.44$ and $x = -5.56$

3.094 Solve $2x^2 - 9x + 3 = 0$.

The quadratic can't be factored. Use the quadratic formula.

Comparing $2x^2 - 9x + 3 = 0$ with $ax^2 + bx + c = 0$, we see that $a = 2$, $b = -9$, $c = 3$.

$$x = \frac{-b \pm \sqrt{b^2 - 4ac}}{2a}$$

$$x = \frac{-(-9) \pm \sqrt{(-9)^2 - 4(2)(3)}}{2(2)}$$

$$x = \frac{9 \pm \sqrt{81-24}}{4}$$

$$x = \frac{9 \pm \sqrt{57}}{4}$$

The two answers are: $x = \frac{9+\sqrt{57}}{4}$ and $x = \frac{9-\sqrt{57}}{4}$

Rounded to two decimal digits, the answers are: $x = 4.14$ and $x = 0.36$

For practice, I'd recommend solving the problems 3.089 to 3.092 by using the quadratic formula as well. The answers would match the ones we obtained by factoring.

PRACTICE PROBLEMS

Solve the following equations by factoring. If factoring is not possible, then use the quadratic formula.

3.095 $x^2 + 14x + 24 = 0$ **3.097** $3x^2 - x - 50 = 2x^2 + 40$ **3.099** $x^2 - 6x - 3 = 0$

3.096 $12x^2 - 4x - 1 = 0$ **3.098** $2x^2 + 19 = 4 - 11x$ **3.100** $5x^2 - 10x + 3 = 0$

EXPONENTIAL EQUATIONS

Exponential equations, as the name suggests, involve exponents. To solve them, *first make the bases the same* (if they aren't the same). Then equate the exponents to form a much simpler equation, and solve.

Consider an example: $2^x = 2^5$ Since we're looking at the same base 2 on both sides, we can conclude that the exponents must be the same on both sides as well. Meaning, x must be 5. So, $x = 5$. Another example: $6^x = 36$ Looks like the bases aren't the same. But 36 can be written as 6^2. So now we have $6^x = 6^2$. Now the bases are the same, so the exponents must be the same as well. Meaning, $x = 2$.

I'd recommend reviewing *exponents* (page 5) and *laws of exponents* (page 43) before you proceed.

3.101 $6^{3x}=6^{14+x}$

The base is 6 on both sides, which means the exponents must also be the same on both sides. Equate the exponents and solve.

$$3x=14+x$$
$${\scriptstyle -x}{\scriptstyle -x}$$
$$2x=14$$
$$\frac{2x}{2}=\frac{14}{2}$$
$$x=7$$

3.102 $(-5)^{2x+1}=(-5)^{23}$

The bases are the same, so the exponents must be the same. Equate the exponents and solve.

$$2x+1=23$$
$${\scriptstyle -1}{\scriptstyle -1}$$
$$2x=22$$
$$\frac{2x}{2}=\frac{22}{2}$$
$$x=11$$

3.103 $2^{x+5}=8$

The bases are not the same. To equate them, notice that 8 can be written as 2^3.

$$2^{x+5}=2^3$$

Now the bases are the same. Equate the exponents and solve.

$$x+5=3$$
$${\scriptstyle -5}{\scriptstyle -5}$$
$$x=-2$$

3.104 $3^{x-7}=\dfrac{1}{3^2}$

The bases are the same, but 3^2 is in the denominator. To bring it to the numerator, make its exponent negative (page 43, law 4 backwards).

$$3^{x-7}=3^{-2}$$

The bases are the same. Equate the exponents.

$$x-7=-2$$
$${\scriptstyle +7}{\scriptstyle +7}$$
$$x=5$$

3.105 $64=8^{6-4x}$

The bases are not the same. To equate them, notice that 64 can be written as 8^2.

$$8^2=8^{6-4x}$$

The bases are the same. Equate the exponents.

$$2=6-4x$$
$${\scriptstyle -6}{\scriptstyle -6}$$
$$-4=-4x$$
$$\frac{-4}{-4}=\frac{-4x}{-4}$$
$$1=x$$

3.106 $\dfrac{1}{3^x}=3^{2x+9}$

The bases are the same, but 3^x is in the denominator. To bring it to the numerator, make its exponent negative (page 43, law 4 backwards).

$$3^{-x}=3^{2x+9}$$

The bases are the same. Equate the exponents.

$$-x=2x+9$$
$${\scriptstyle -2x}{\scriptstyle -2x}$$
$$-3x=9$$
$$\frac{-3x}{-3}=\frac{9}{-3}$$
$$x=-3$$

PRACTICE PROBLEMS

3.107 $2^{5x}=2^{12-x}$

3.108 $4^{5x}=2^{x+9}$

3.109 $3^{x+8}=27$

3.110 $25^3=5^{2x-2}$

3.111 $16^4=4^{5x-2}$

3.112 $7^{2x+4}=49$

RADICAL EQUATIONS

Recall that a radical means a root: not a beet root or a carrot, but a square root, cube root, fourth root etc. of a number. This being a typical student's first book on algebra, we will limit our discussion to square roots only. We will also consider examples that are not overly complicated. Before we get started, evaluate your knowledge of radicals. If need be, please visit page 18 for a refresher.

Consider a very elementary radical equation:

$$\sqrt{9}=3$$

We're only looking at the principal root (positive root). I know this equation doesn't have a variable, but please allow me to make my point.

Squaring both sides, we'd get:

$$(\sqrt{9})^2=3^2$$
$$\sqrt{9}\cdot\sqrt{9}=3\cdot3$$
$$\sqrt{81}=9$$
$$9=9$$

From start to finish, what happened was that the radical sign $(\sqrt{})$ got eliminated and the 9 from under it was set free. How did we achieve that? By simply squaring both sides! Now imagine if, instead of 9, x were under the radical sign. We'd use the same method to set it free so we could solve for x.

When solving radical equations, first isolate the radical. Then square both sides of the equation to eliminate the radical sign. Then solve for the variable.

3.113 $\sqrt{x}=4$

Square both sides of the equation to eliminate the radical sign.

$$(\sqrt{x})^2=4^2$$
$$x=4\cdot4$$
$$x=16$$

3.114 $\sqrt{2x-5}=7$

Square both sides of the equation to eliminate the radical sign.

$$(\sqrt{2x-5})^2=7^2$$
$$2x-5=7\cdot7$$
$$2x-5=49$$
$${}^{+5}\qquad{}^{+5}$$
$$2x=54$$
$$\frac{2x}{2}=\frac{54}{2}$$
$$x=27$$

3.115 $1+\sqrt{8+4x}=5$

First, isolate the radical.

$$1+\sqrt{8+4x}=5$$
$$^{-1}\phantom{+\sqrt{8+4x}}{}^{-1}$$
$$\sqrt{8+4x}=4$$

Now square both sides to eliminate the radical sign.

$$(\sqrt{8+4x})^2=4^2$$
$$8+4x=4\cdot4$$
$$8+4x=16$$
$$^{-8}{}^{-8}$$
$$4x=8$$
$$\frac{4x}{4}=\frac{8}{4}$$
$$x=2$$

3.116 $3\sqrt{6x-12}+19=6+7$

This equation looks like scrambled eggs spilled on the floor! Everything is all over the place. Let's first simplify it and get the radical by itself.

$$3\sqrt{6x-12}+19=6+7$$
$$3\sqrt{6x-12}+19=13$$
$$\phantom{3\sqrt{6x-12}+19}{}_{-19}\quad{}_{-19}$$
$$3\sqrt{6x-12}=-6$$
$$\frac{3\sqrt{6x-12}}{3}=\frac{-6}{3}$$
$$\sqrt{6x-12}=-2$$

Now square both sides to eliminate the radical sign.

$$\left(\sqrt{6x-12}\right)^2=(-2)^2$$
$$6x-12=(-2)\cdot(-2)$$
$$6x-12=4$$
$${}_{+12}\quad{}_{+12}$$
$$6x=16$$
$$\frac{6x}{6}=\frac{16}{6}$$
$$x=\frac{16}{6}=\frac{8}{3}=2\frac{2}{3}$$

3.117 $2\sqrt{9x+40}-5=17$

First, get the radical by itself.

$$2\sqrt{9x+40}-5=17$$
$$\phantom{2\sqrt{9x+40}-5}{}_{+5}\quad{}_{+5}$$
$$2\sqrt{9x+40}=22$$
$$\frac{2\sqrt{9x+40}}{2}=\frac{22}{2}$$
$$\sqrt{9x+40}=11$$

Now square both sides of the equation.

$$\left(\sqrt{9x+40}\right)^2=11^2$$
$$9x+40=11\cdot11$$
$$9x+40=121$$
$${}_{-40}\quad{}_{-40}$$
$$9x=81$$
$$\frac{9x}{9}=\frac{81}{9}$$
$$x=9$$

PRACTICE PROBLEMS

3.118 $\sqrt{x+8}=-3$

3.119 $5\sqrt{3x}=30$

3.120 $10+\sqrt{5x+45}=20$

3.121 $25+2\sqrt{80+4x}=49$

CHAPTER EXERCISES

3.122 Solve: $2(x+4)-1=11$

3.123 Solve: $1-\frac{8}{3x}=\frac{1}{3x}+\frac{2}{3}$

3.124 Two less than the product of a number and 7 is the same as the sum of the number and 10. What is the number?

3.125 Solve: $\frac{4}{5c}=\frac{2}{15}$

3.126 The ratio of cruisers to choppers at a motorcycle rally is 3:7. If there are 93 cruisers at the rally, then how many choppers are there?

3.127 Solve $3a(b-c)=15b$ for b.

3.128 Solve $\frac{D}{k}+16=1-c$ for k.

3.129 Solve the system of equations: $x=-y+5$ and $6x+8y=34$.

3.130 Hillary is 3 inches taller than Donald. If the sum of their heights is 51 inches, then how tall is each of them?

3.131 Solve the system of equations: $2x+4y=6$ and $3x-y=16$.

3.132 Two eggs and 3 tangerines together weigh 13 ounces. If 3 eggs and 8 tangerines together weigh 30 ounces, then how much does each egg and each tangerine weigh?

3.133 Solve: $|x-8|+3=7$

3.134 Solve: $14+|2-5x|=6$

3.135 Solve by factoring: $3x^2-2x+4=14-x$

3.136 Use the quadratic formula to solve $2x^2-5x-6=0$. Round your answers to two decimal places.

3.137 Solve: $16^{x+5}=2^{6x-10}$

3.138 Solve: $\sqrt{2x-7}-13=2$

CHAPTER 4
FUNCTIONS AND GRAPHS

MEANING OF A FUNCTION

Imagine you drive into a gas station to fuel your thirsty car. The more gas you pump, the more money you will pay. Let's say gas costs $2.50 a gallon. For 5 gallons of gas, you will pay $2.50 \times 5 = \$12.50$; for 10 gallons, you shall pay $2.50 \times 10 = \$25$; for x gallons of gas, thou shalt pay $2.50 \times x$ dollars. Clearly, the amount of money you spend would *depend on* the amount of gas you pump. In function terminology, money (M) would be the *dependent variable* because it would depend on the amount of gas, whereas the amount of gas (x) would be the *independent variable*.

In terms of an equation, we can write $M = 2.50x$. However, in function notation, the equation becomes $M(x) = 2.50x$. The meaning of $M(x)$ is not "M times x" but "M of x" or "M as a function of x." Similarly, $f(x)$ is read "f of x"; $g(x)$ is read "g of x". I hope you get the point.

In simple language, a **function** is an equation that shows the relationship between the dependent and independent variables.

EVALUATING FUNCTIONS

Imagine you're given the function $M(x) = 2.50x$ and you're asked to find the value of the function when $x = 16$. Meaning, how much money would you spend if you pumped 16 gallons of gas? In function notation, you're asked to find $M(16)$ which is the value of $M(x)$ when $x = 16$. That would be $M(16) = 2.50 \times 16 = 40 = \$40$. What about $M(9)$? That would be $M(9) = 2.50 \times 9 = 22.5 = \$22.50$.

Consider another function, $f(x) = 2x + 5$.

What is $f(4)$? That would be $f(4) = 2 \cdot (4) + 5 = 8 + 5 = 13$

What is $f(0)$? That would be $f(0) = 2 \cdot (0) + 5 = 0 + 5 = 5$

What is $f(-3)$? That would be $f(-3) = 2 \cdot (-3) + 5 = -6 + 5 = -1$

What is $f(c)$? That would be $f(c) = 2 \cdot (c) + 5 = 2c + 5$

Do you see the pattern? We're simply replacing x with the given value for which the function is to be evaluated, whether it is 4, 0, –3, or c.

The topic of functions is quite vast. But in this book we will learn the basic mechanics of functions and how to graph a few core types of functions using one method in each case.

PRACTICE PROBLEMS

4.001 Evaluate the function $f(x)=3x-8$ for $x=4$, $x=-2$ and $x=0$.

4.002 Given the function $g(x)=x^2+7$, find $g(6)$, $g(-10)$ and $g(1)$.

4.003 If $F(x)=\dfrac{x+4}{x-6}$ then evaluate $F(16)$ and $F(22)$.

THE COORDINATE PLANE

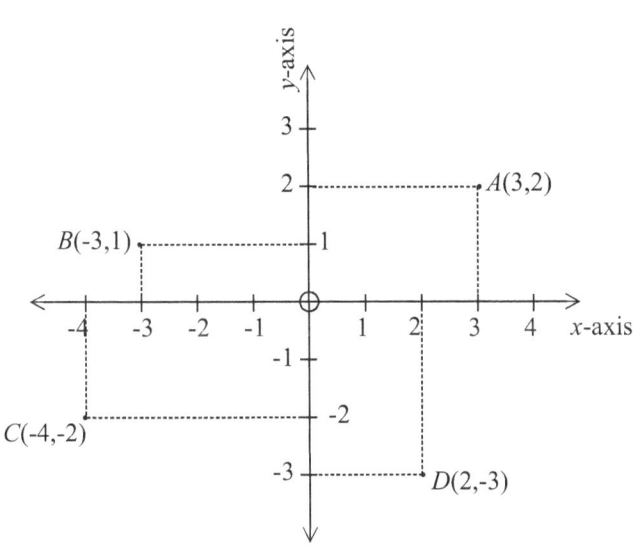

The coordinate plane consists of a horizontal number line (the **x-axis**) and a vertical number line (the **y-axis**). The two axes intersect at their 0, shown as a circle in the figure on the left. This point of intersection is called the **origin**.

Any point plotted on the coordinate plane can be expressed as an **ordered pair** (x,y) where x is the x-coordinate (horizontal distance from the origin), and y is the y-coordinate (vertical distance from the origin). For example, to plot the point $A(3,2)$, start at the origin and slide right 3 units and slide up 2 units. To plot the point $B(-3,1)$, slide left 3 units from the origin and up 1 unit. In a similar fashion, points C and D can be plotted as well.

Later in this chapter, we will study how to plot the graphs of various functions on the coordinate plane.

DOMAIN AND RANGE

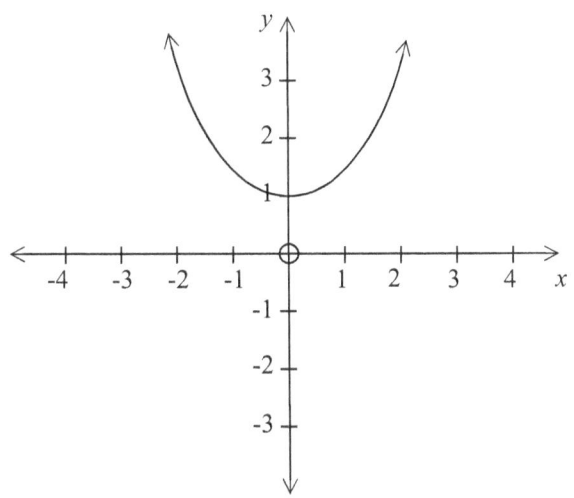

Figure on the left shows the graph of a quadratic function (about which we'll study later in this chapter). The curve extends forever on both sides. Meaning, it keeps rising and also keeps growing horizontally left and right. That's what the arrows mean.

Domain comprises all the possible x-values (on the horizontal axis) the curve will trace. Meaning, how far left and right will the curve span? Well, that looks like negative infinity to positive infinity, or all the possible real numbers. We therefore state the domain as *all real numbers*, denoted by $\mathbb{R}$. It's just a fancy looking R, nothing more.

Range comprises all the possible y-values (on the vertical axis) the curve will trace. Meaning, how far up and down will the curve reach? Notice that the curve drops until it touches 1, and rises all the way to infinity. Therefore, the range will be from 1 (inclusive) to infinity. Infinity is a concept, not a specific number, and is considered unreachable and therefore not inclusive. It is denoted by ∞. We state the range as the interval $[1,\infty)$. Smaller value first. Notice the *bracket* on the left and *parenthesis* on the right? When a number is *included*, we use a bracket. When a number is *not included*, we use a parenthesis.

Now consider the graph of an exponential function (about which we'll study later in this chapter) as shown on the right. The curve extends left and right forever. On the left, the curve dips closer and closer to the x-axis (horizontal axis) but never touches it. On the right, the curve reaches for the stars!

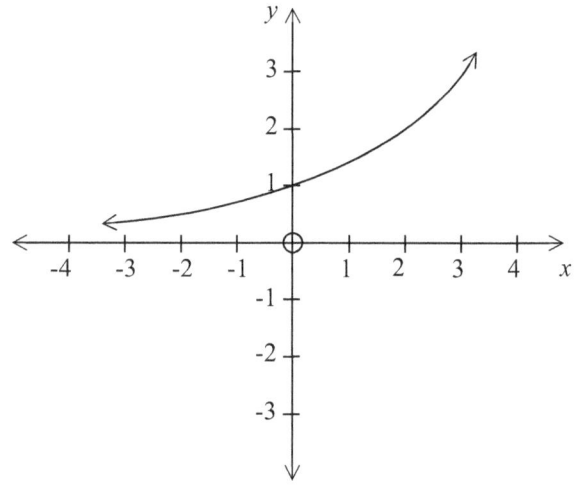

What is the domain (the possible x-values) of this function? Since it grows left and right forever, it is *all real numbers*, $\mathbb{R}$. What is the range (the possible y-values) of this function? Since the curve never quite drops to zero, and rises all the way to infinity, the range would be zero (not inclusive) to positive infinity (not inclusive), written as the interval $(0,\infty)$.

Note: If you want, you can write *all real numbers*, $\mathbb{R}$, as the interval $(-\infty,\infty)$. Smaller value first.

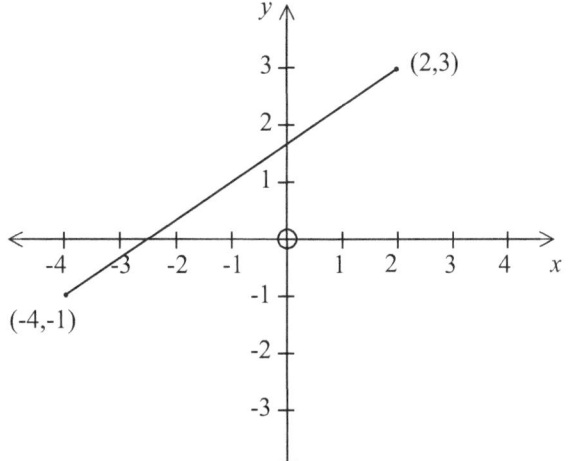

Consider a third example, a line segment (part of a line) as shown on the left. We'll study how to plot lines starting the next section, *ha!* The line segment, as you can tell, has definite end points.

What appears to be the domain of this function? Well, it reaches leftward until the x-coordinate -4 (inclusive), and rightward until 2 (inclusive). So the domain is the interval $[-4,2]$. The brackets mean both -4 and 2 are *included*. What is the range? The line segment drops up to the y-coordinate -1 and rises up to 3. So the range must be the interval $[-1,3]$.

Domain is about the x-values. Range is about the y-values. How to remember? In the alphabet, D comes before R. In the alphabet, x comes before y.

LINEAR FUNCTIONS

A **linear function** represents a line that extends infinitely in both directions. The picture above shows a line *segment*, but now we will consider the entire line. A linear function is of the form $f(x)=ax+b$, where a is the slope, and b is the y-intercept. Slope is the measure of *steepness* of a line. Slope is calculated as $\dfrac{\text{rise}}{\text{run}}$ and is positive if the line is going uphill from left to right, or negative if the line is going downhill from left to right. y-intercept is the point where the line cuts the y-axis (vertical axis).

Rise is the difference between the *y*-coordinates of two points on the line. *Run* is the difference between the *x*-coordinates of those same two points.

When plotting on a coordinate plane, the $f(x)$ or $g(x)$ or $h(x)$ etc. represents the *y*-values in any function (linear or otherwise). So first write $f(x)=ax+b$ as $y=ax+b$.

Consider a sample function $f(x)=\frac{1}{3}x+2$ which is essentially $y=\frac{1}{3}x+2$. We know that $y=ax+b$ means $y=\left(\frac{\text{rise}}{\text{run}}\right)x+(y\text{-intercept})$. In this example, what is the *run*? It's 3. Go those many steps to the left and right of zero on the *x*-axis (horizontal axis). You'll reach *x*-coordinates −3 and +3. Calculate the *y*-coordinates at $x=-3$, +3 and 0 using the equation you have. Then mark those points and draw a line through them. Table below shows computations for the *y*-coordinates. On the right is the graph.

x	$y=\frac{1}{3}x+2$
−3	$y=\frac{1}{3}\cdot(-3)+2=\frac{-3}{3}+2=-1+2=1$
0	$y=\frac{1}{3}\cdot(0)+2=0+2=2$
3	$y=\frac{1}{3}\cdot(3)+2=\frac{3}{3}+2=1+2=3$

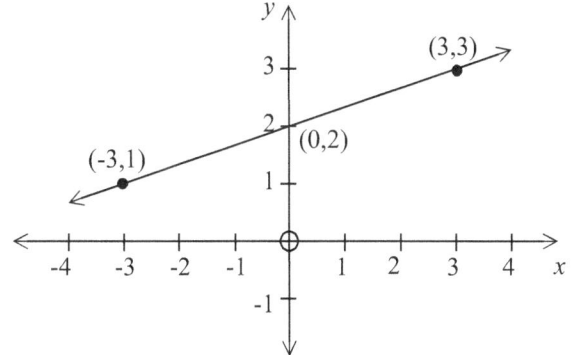

Three points are enough to graph a line. We chose the above three convenient *x*-values for ease of calculations. You can choose any *x*-values you'd like and plot as many points as you want.

Notice that the line grows forever on the left and right (horizontally) which means the domain must be *all real numbers*, $\mathbb{R}$, or the interval $(-\infty,\infty)$. It also forever drops on the left and rises on the right (vertically) which means the range must also be *all real numbers*. In general, for any sloping line, the domain and range are always $\mathbb{R}$.

4.004 Graph the linear function $f(x)=\frac{2}{3}x+1$. What are its domain and range?

The function $f(x)=\frac{2}{3}x+1$ means $y=\frac{2}{3}x+1$. What is the *run*? It's 3. So go 3 paces to the left and right of 0 on the *x*-axis. You'll find *x*-coordinates −3 and +3. Calculate the *y*-coordinates at $x=-3$, +3 and 0 using the equation you have. Then plot those points and draw a line through them. Table on the next page shows the calculations. On the right of it is the graph.

The domain and range of the linear function are both $\mathbb{R}$, or the interval $(-\infty,\infty)$.

x	$y=\dfrac{2}{3}x+1$
-3	$y=\dfrac{2}{3}\cdot(-3)+1=\dfrac{-6}{3}+1=-2+1=-1$
0	$y=\dfrac{2}{3}\cdot(0)+1=0+1=1$
3	$y=\dfrac{2}{3}\cdot(3)+1=\dfrac{6}{3}+1=2+1=3$

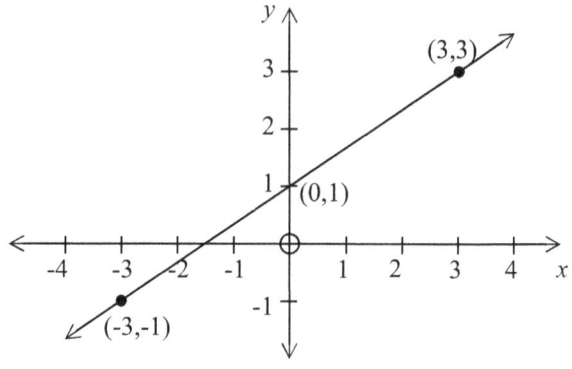

4.005 Graph the function $y=\dfrac{-1}{2}x+4$ and state its domain and range.

What is the *run*? It's 2. So go 2 paces to the left and right of 0 on the *x*-axis. You'll find the *x*-coordinates –2 and +2. Calculate the *y*-coordinates at $x=-2$, $+2$ and 0 using the equation you have. Then plot those points and draw a line through them. Table below shows the calculations. The graph is on its right. The domain and range of the function are both $\mathbb{R}$, or the interval $(-\infty,\infty)$.

x	$y=\dfrac{-1}{2}x+4$
-2	$y=\dfrac{-1}{2}\cdot(-2)+4=\dfrac{2}{2}+4=1+4=5$
0	$y=\dfrac{-1}{2}\cdot(0)+4=0+4=4$
2	$y=\dfrac{-1}{2}\cdot(2)+4=\dfrac{-2}{2}+4=-1+4=3$

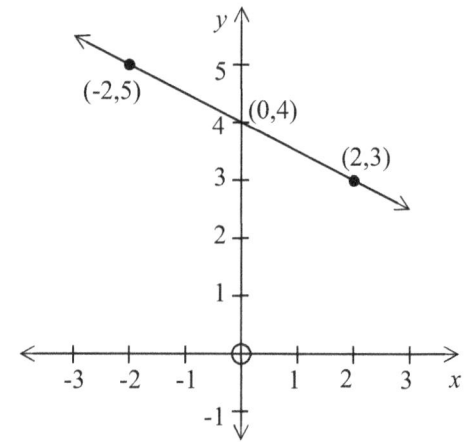

4.006 Graph $2x+y=5$ and write its domain and range.

The format of this linear function is not in the form $y=ax+b$. So, first solve it for y. Refer page 47 if you need a refresher on how to solve for a variable.

$$2x+y=5$$
$$\underset{-2x}{}\quad\underset{-2x}{}$$
$$y=5-2x$$
$$y=-2x+5$$
$$y=\dfrac{-2}{1}x+5$$

What is the *run*? It's 1. So go 1 step to the left and right of 0 on the *x*-axis. That's *x*-coordinates –1 and +1. Calculate the *y*-coordinates at $x=-1$, $+1$ and 0 using the simpler form $y=-2x+5$. Then plot those points and draw a line through them. Table at the top of the next page shows the calculations. Adjacent to it is the graph of the linear function. Notice that the scale on the *y*-axis is different from the scale on the *x*-axis. It's okay to do that if you're trying to draw a tall/wide graph and have limited space. The domain and range of the function are both $\mathbb{R}$, or $(-\infty,\infty)$.

x	$y=-2x+5$
-1	$y=-2\cdot(-1)+5=2+5=7$
0	$y=-2\cdot(0)+5=0+5=5$
1	$y=-2\cdot(1)+5=-2+5=3$

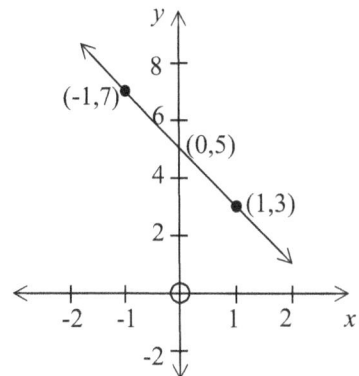

4.007 Graph $h(x)=\dfrac{x}{5}-3$ and state its domain and range.

Notice that $\dfrac{x}{5}$ is the same as $\dfrac{1}{5}x$. The linear function is thus $y=\dfrac{1}{5}x-3$ where the *run* is 5. So go 5 paces to the left and right of 0 on the *x*-axis. That's *x*-coordinates -5 and $+5$. Calculate the *y*-coordinates at $x=-5$, $+5$ and 0 using the equation you have. Then plot those points and draw a line through them. Table below shows the computations. On its right is the graph. The domain and range of the function are both $\mathbb{R}$, or $(-\infty,\infty)$.

x	$y=\dfrac{1}{5}x-3$
-5	$y=\dfrac{1}{5}\cdot(-5)-3=\dfrac{-5}{5}-3=-1-3=-4$
0	$y=\dfrac{1}{5}\cdot(0)-3=0-3=-3$
5	$y=\dfrac{1}{5}\cdot(5)-3=\dfrac{5}{5}-3=1-3=-2$

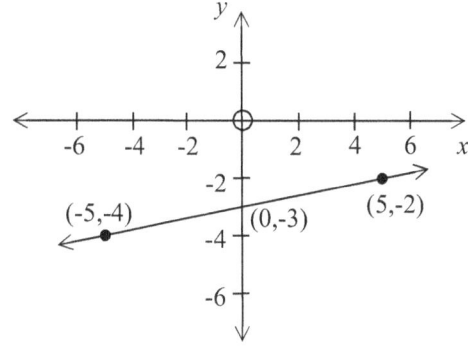

4.008 Graph $p(x)=4x$ and state its domain and range.

The function $p(x)=4x$ can be written as $y=4x$ or more precisely as $y=\dfrac{4}{1}x+0$. The *run* is 1. So go 1 step to the left and right of 0 on the *x*-axis. That's *x*-coordinates -1 and $+1$. Calculate the *y*-coordinates at $x=-1$, $+1$ and 0 using the simpler form $y=4x$. Then mark those points and draw a line through them. Table below shows the calculations. Figure on its right shows the graph. The domain and range of the function are both $\mathbb{R}$, or $(-\infty,\infty)$.

x	$y=4x$
-1	$y=4\cdot(-1)=-4$
0	$y=4\cdot(0)=0$
1	$y=4\cdot(1)=4$

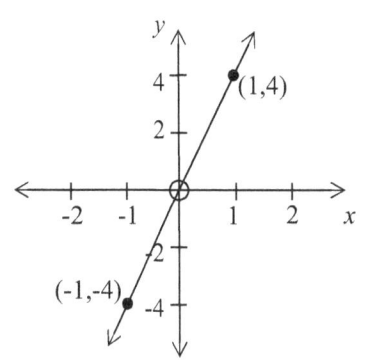

4.009 Graph $x=3y-6$ and state its domain and range.

The format of this linear function is not in the form $y=ax+b$. So, first solve it for y.

$$x \underset{-3y \quad -3y}{=3y-6}$$

$$x-3y=-6$$
$$\underset{-x \qquad -x}{}$$

$$-3y=-6-x$$

$$\frac{-3y}{-3}=\frac{-6-x}{-3}$$

$$y=\frac{-6}{-3}-\frac{x}{-3}$$

$$y=2+\frac{x}{3}$$

$$y=\frac{1}{3}x+2$$

What is the *run*? It's 3. So go 3 paces to the left and right of 0 on the *x*-axis. That's *x*-coordinates −3 and +3. Calculate the *y*-coordinates at $x=-3$, +3 and 0 using the equation you have. Then plot those points and draw a line through them. Table below shows the calculations. The graph is on its right. The domain and range of the function are both $\mathbb{R}$, or $(-\infty,\infty)$.

x	$y=\frac{1}{3}x+2$
−3	$y=\frac{1}{3}\cdot(-3)+2=\frac{-3}{3}+2=-1+2=1$
0	$y=\frac{1}{3}\cdot(0)+2=0+2=2$
3	$y=\frac{1}{3}\cdot(3)+2=\frac{3}{3}+2=1+2=3$

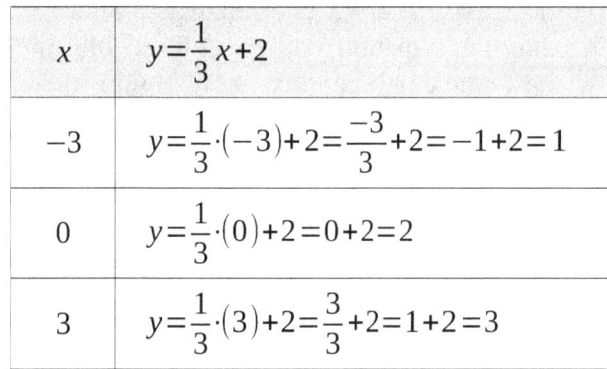

PRACTICE PROBLEMS
Graph the following linear functions:

4.010 $3x+5y=30$

4.011 $y=\frac{2}{7}x-5$

4.012 $f(x)=\frac{x}{4}+2$

4.013 $g(x)=-\frac{3}{4}x-3$

4.014 $x=\frac{1}{5}y+2$

4.015 $q(x)=-5x$

EXPONENTIAL FUNCTIONS

An **exponential function** represents a curve that turns away from a horizontal line at an ever increasing pace on one side, and approaches the same line at an ever slowing pace on the other side. This line that the curve approaches but never touches is called an **asymptote**. The function is of the form $f(x)=a\cdot b^x$ or $y=a\cdot b^x$ where a is the *y*-intercept and b is the factor of growth (or decay). a can be positive or negative, but b has to be positive and not equal to 1. Figure on the right shows four shapes of exponential functions depending on their a and b values.

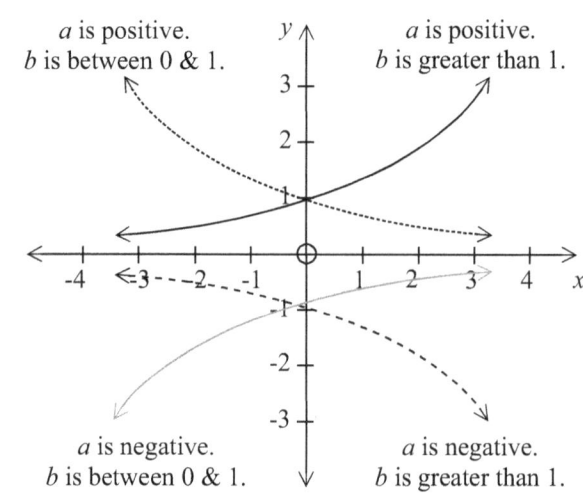

a is positive. b is between 0 & 1.
a is positive. b is greater than 1.
a is negative. b is between 0 & 1.
a is negative. b is greater than 1.

Notice that an exponential function always curves *away* from the asymptote (here, the *x*-axis). Since it grows left and right forever, the domain is *all real numbers*, $\mathbb{R}$. In the vertical direction, the curve approaches but never reaches zero at one end, and heads toward either the stars or the abyss at the other end. Thus, the range would be zero (not inclusive) to either positive infinity or negative infinity (not inclusive), written as the interval $(0,\infty)$ or $(-\infty,0)$. Smaller value first.

When it comes to graphing exponential functions, plot three points and draw a smooth curve through them. For ease of calculations, always use the *x*-coordinates of the three points as -1, $+1$, and 0. Plug these *x*-coordinates into the given exponential function to find the *y*-coordinates.

To get an idea of how it works, let's jump into examples. Since we will use the *laws of exponents* in our calculations, please review them from page 43 if you need to before proceeding.

4.016 Graph the exponential function $f(x)=2\cdot(3)^x$ and state its domain and range.

First calculate the *y*-coordinates at $x=-1$, $+1$, and 0 using the function which may be written as $y=2\cdot(3)^x$ Then plot the points and draw a smooth curve through them. Table below shows the calculations. On its right is the graph. The domain of the function is $\mathbb{R}$, or the interval $(-\infty,\infty)$. The range of the function is the interval $(0,\infty)$.

x	$y=2\cdot(3)^x$
-1	$y=2\cdot(3)^{-1}=2\cdot\left(\dfrac{1}{3}\right)=\dfrac{2}{3}=0.\overline{6}$
0	$y=2\cdot(3)^0=2\cdot(1)=2$
1	$y=2\cdot(3)^1=2\cdot(3)=6$

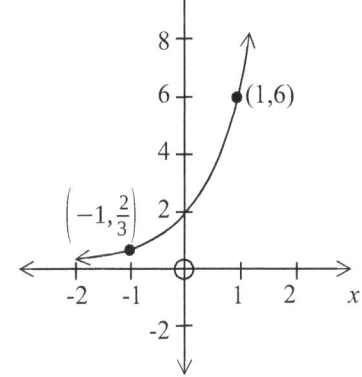

4.017 Graph the function $y=125\cdot(5)^{x-3}$ and write its domain and range.

The function is not in the form $y=a\cdot b^x$.
So first simplify it.

$y=125\cdot(5)^{x-3}$

$y=125\cdot\dfrac{5^x}{5^3}$ Refer page 43, law 2.

$y=125\cdot\dfrac{5^x}{125}$

$y=\dfrac{125\cdot(5)^x}{125}$

$y=\dfrac{1\cdot(5)^x}{1}$

$y=5^x$

x	$y=5^x$
-1	$y=5^{-1}=\dfrac{1}{5}=0.2$
0	$y=5^0=1$
1	$y=5^1=5$

Now calculate the *y*-coordinates at $x=-1$, $+1$, and 0. Then plot the points and draw a smooth curve through them. Table and graph are on the right. The domain of the function is $\mathbb{R}$, or $(-\infty,\infty)$, and the range is $(0,\infty)$.

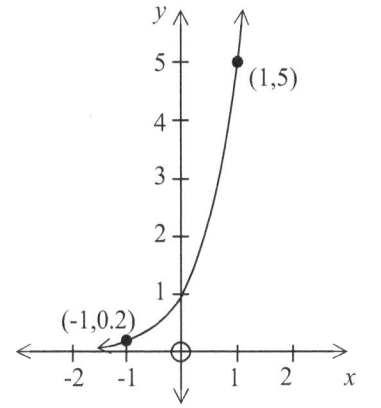

4.018 Graph $y=2\cdot(4)^{-x}$ and mention its domain and range.

The function is not in the form $y=a\cdot b^x$.
So first simplify it.

$$y=2\cdot(4)^{-x}$$

$$y=2\cdot\frac{1}{4^x} \qquad \text{Refer page 43, law 4.}$$

$$y=2\cdot\frac{1^x}{4^x} \qquad \text{because } 1^x=1.$$

$$y=2\cdot\left(\frac{1}{4}\right)^x \qquad \text{Refer page 43, law 8.}$$

$$y=2\cdot(0.25)^x$$

Now calculate the y-coordinates at $x=-1$, $+1$, and 0. Then plot the points and draw a smooth curve through them. Table and graph are on the right. The domain of the function is $\mathbb{R}$, or $(-\infty,\infty)$, and the range is $(0,\infty)$.

x	$y=2\cdot(0.25)^x$
-1	$y=2\cdot(0.25)^{-1}=2\cdot\dfrac{1}{0.25}=2\cdot4=8$
0	$y=2\cdot(0.25)^0=2\cdot1=2$
1	$y=2\cdot(0.25)^1=2\cdot(0.25)=0.5$

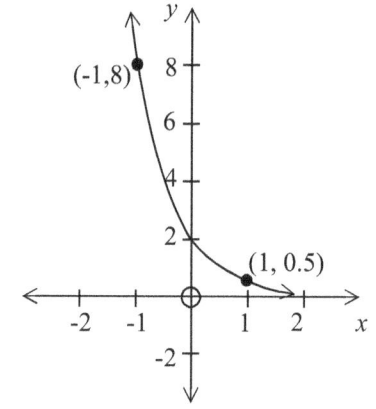

4.019 Graph $h(x)=-\dfrac{4}{9}\cdot3^{x+2}$ and state its domain and range.

The function is not in the form $y=a\cdot b^x$.
So first simplify it.

$$y=-\frac{4}{9}\cdot3^{x+2}$$

$$y=-\frac{4}{9}\cdot3^x\cdot3^2 \qquad \text{Refer page 43, law 1.}$$

$$y=-\frac{4}{9}\cdot3^x\cdot9$$

$$y=-\frac{36}{9}\cdot3^x$$

$$y=-4\cdot(3)^x$$

Now calculate the y-coordinates at $x=-1$, $+1$, and 0. Then plot the points and draw a smooth curve through them. Table and graph are on the right. The domain of the function is $\mathbb{R}$, or $(-\infty,\infty)$, and the range is $(-\infty,0)$. Smaller value first.

x	$y=-4\cdot(3)^x$
-1	$y=-4\cdot(3)^{-1}=-4\cdot\dfrac{1}{3}=-\dfrac{4}{3}=-1.\bar{3}$
0	$y=-4\cdot(3)^0=-4\cdot1=-4$
1	$y=-4\cdot(3)^1=-4\cdot3=-12$

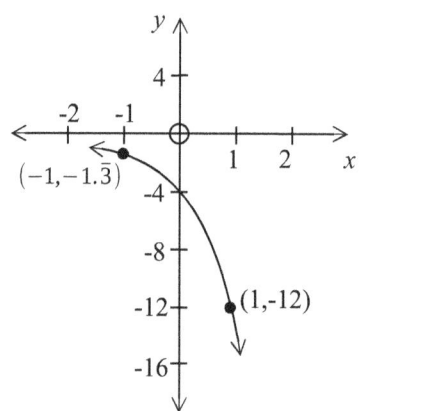

4.020 Graph $p(x)=-4\cdot\left(\dfrac{1}{2}\right)^{x}$ and write its domain and range.

First calculate the y-coordinates at $x=-1$, $+1$, and 0 using the function which may be written as $y=-4\cdot(0.5)^{x}$ Then plot the points and draw a smooth curve through them. Table and graph are shown below. The domain of the function is $\mathbb{R}$, or $(-\infty,\infty)$, and the range is $(-\infty,0)$.

x	$y=-4\cdot(0.5)^{x}$
-1	$y=-4\cdot(0.5)^{-1}=-4\cdot\dfrac{1}{0.5}=-4\cdot2=-8$
0	$y=-4\cdot(0.5)^{0}=-4\cdot1=-4$
1	$y=-4\cdot(0.5)^{1}=-4\cdot(0.5)=-2$

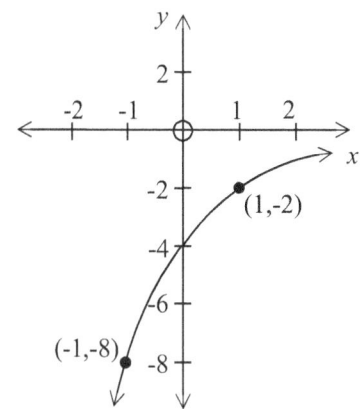

PRACTICE PROBLEMS

Graph the following exponential functions and mention their domain and range:

4.021 $y=6\cdot\left(\dfrac{1}{3}\right)^{x}$

4.022 $f(x)=-15\cdot(3)^{x-1}$

4.023 $y=-(4)^{x}$

4.024 $q(x)=-8\cdot(4)^{-x}$

4.025 $y=4\cdot(5)^{x+2}$

QUADRATIC FUNCTIONS

A **quadratic function** represents a *parabola*: a U-shaped curve that opens either upward or downward. Accordingly, at its center lies either the lowermost point or the uppermost point known as the **vertex**. The two arms of a parabola are mirror images of each other on either side of the vertex. Figure on the right shows two sample configurations of a parabola.

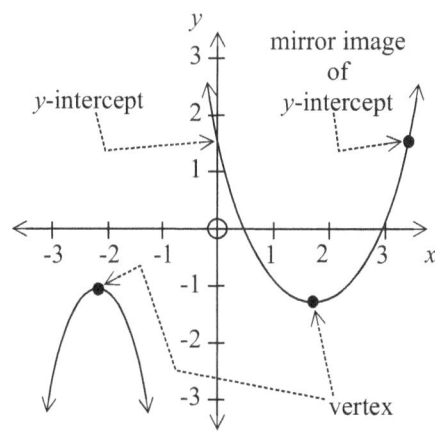

A quadratic function is of the standard form $f(x)=ax^{2}+bx+c$ or $y=ax^{2}+bx+c$, where c is the y-intercept, and a and b can be used to find the location of the vertex.

To graph a quadratic function, first make sure it is in the standard form, $y=ax^{2}+bx+c$. Calculate the x-coordinate of the vertex, which is $\dfrac{-b}{2a}$. Find the y-coordinate by plugging the x-coordinate into the standard form. Once you have placed the vertex, mark the y-intercept at $(0,c)$ and its mirror image on the *other side* of the vertex. The parabola is a smooth curve through these three points, with its arms growing endlessly. Domain (horizontal reach) of a quadratic function is always $\mathbb{R}$, or $(-\infty,\infty)$, whereas range (vertical reach) is from the y-coordinate of vertex (included) to either positive or negative infinity.

The examples starting the next page demonstrate this procedure.

4.026 Graph the quadratic function $y=2x^2+8x+5$ and state its domain and range.

Comparing the function with $y=ax^2+bx+c$, note that $a=2$, $b=8$, $c=5$.

The x-coordinate of the vertex is $\dfrac{-b}{2a}=\dfrac{-8}{2\cdot2}=\dfrac{-8}{4}=-2$

To find the y-coordinate of the vertex, plug -2 for x into the quadratic function:

$y=2(-2)^2+8(-2)+5$
$y=2(4)-16+5$
$y=8-16+5=-3$

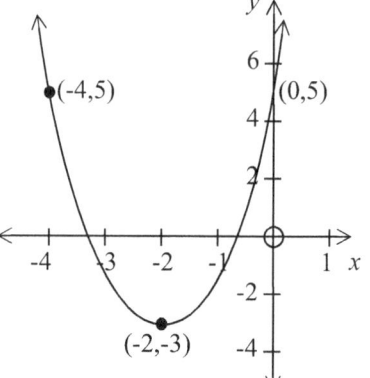

Mark the vertex at $(-2,-3)$ and the y-intercept at $(0,c)$ which is $(0,5)$. The mirror image of the y-intercept will be horizontally twice as far as the vertex, and so its x-coordinate will be -4. But its y-coordinate will be the same as that of the y-intercept, 5. Mark the mirror point at $(-4,5)$. On the right is drawn the parabola with its domain $(-\infty,\infty)$. Its range would be $[-3,\infty)$ where -3 is included because the parabola drops as low as -3.

4.027 Graph the quadratic function $f(x)=(x-3)^2+1$ and write its domain and range.

The function is not in the standard form $y=ax^2+bx+c$. So first simplify it.

$y=(x-3)^2+1$
$y=(x-3)(x-3)\ +1$
$y=x^2-3x-3x+9\ +1$
$y=x^2-6x+10$

Comparing with the standard form, note that $a=1$, $b=-6$, $c=10$.

The x-coordinate of the vertex is $\dfrac{-b}{2a}=\dfrac{-(-6)}{2\cdot1}=\dfrac{6}{2}=3$.

To find the y-coordinate of the vertex, plug 3 for x into the quadratic function. You may use either the standard form or the original function. You'll get the same answer.

$y=3^2-6(3)+10$
$y=9-18+10=1$

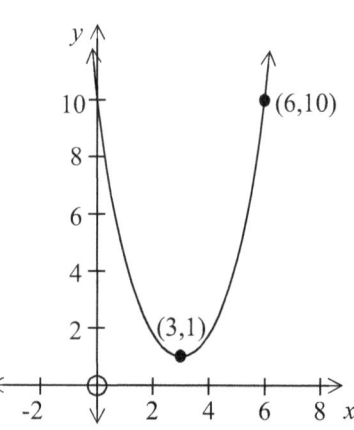

Mark the vertex at $(3,1)$ and the y-intercept at $(0,c)$ which is $(0,10)$. The mirror image of the y-intercept will be horizontally twice as far as the vertex, and so its x-coordinate will be 6. But its y-coordinate will be the same as that of the y-intercept, 10. Mark the mirror point at $(6,10)$. Draw the parabola through those points. Its domain is $(-\infty,\infty)$ and its range is $[1,\infty)$.

4.028 Graph the function $y=(x+2)(x-6)$ and state its domain and range.

The function is not in the standard form $y=ax^2+bx+c$. So first simplify it.
$y=(x+2)(x-6)$
$y=x^2-6x+2x-12$
$y=x^2-4x-12$

Comparing with the standard form, note that $a=1$, $b=-4$, $c=-12$.

The x-coordinate of the vertex is $\dfrac{-b}{2a}=\dfrac{-(-4)}{2\cdot 1}=\dfrac{4}{2}=2$.

To find the y-coordinate of the vertex, plug 2 for x into the quadratic function. You may use either the standard form or the original function. You'll get the same answer.

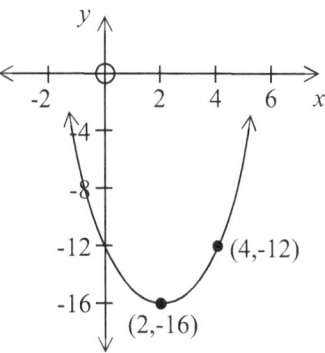

$$y=2^2-4(2)-12$$
$$y=4-8-12=-16$$

Mark the vertex at $(2,-16)$ and the y-intercept at $(0,c)$ which is $(0,-12)$. The mirror image of the y-intercept will be horizontally twice as far as the vertex, and so its x-coordinate will be 4. But its y-coordinate will be the same as that of the y-intercept, -12. Mark the mirror point at $(4,-12)$. Draw the parabolic curve. Its domain is $(-\infty,\infty)$ and its range is $[-16,\infty)$.

4.029 Graph the function $g(x)=-5x^2+30x$ and mention its domain and range.

The function may be written as $y=-5x^2+30x+0$. Comparing it with $y=ax^2+bx+c$, note that $a=-5$, $b=30$, $c=0$.

The x-coordinate of the vertex is $\dfrac{-b}{2a}=\dfrac{-30}{2\cdot(-5)}=\dfrac{-30}{-10}=3$.

To find the y-coordinate of the vertex, plug 3 for x into the quadratic function:

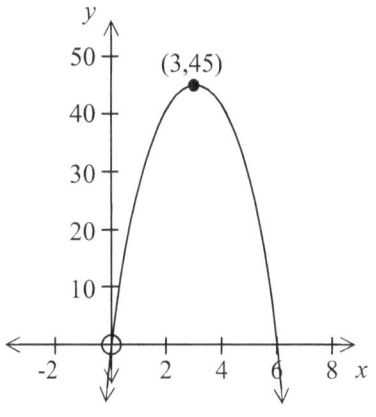

$$y=-5(3)^2+30(3)+0$$
$$y=-5\cdot 9+90$$
$$y=-45+90=45$$

Mark the vertex at $(3,45)$ and the y-intercept at $(0,c)$ which is $(0,0)$. The mirror image of the y-intercept will be horizontally twice as far as the vertex, and so its x-coordinate will be 6. But its y-coordinate will be the same as that of the y-intercept, 0. Mark the mirror point at $(6,0)$. On the right is drawn the parabola with domain $(-\infty,\infty)$ and range $(-\infty,45]$. Remember, smaller value is always written first in an interval.

4.030 Graph the quadratic function $y=-3x^2$ and write its domain and range.

Comparing the function with $y=ax^2+bx+c$, note that $a=-3$, $b=0$, $c=0$.

The x-coordinate of the vertex is $\dfrac{-b}{2a}=\dfrac{-(0)}{2\cdot(-3)}=\dfrac{0}{-6}=0$.

To find the y-coordinate of the vertex, plug 0 for x into the function:

$$y=-3(0)^2=-3(0)=0$$

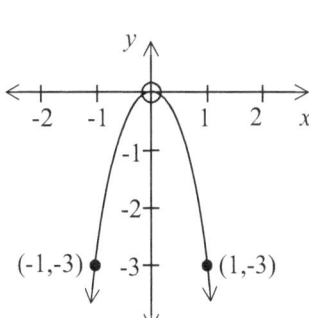

The vertex is at $(0,0)$ and so is the y-intercept. In such cases, find the y-coordinates of two neighboring points, say at $x=-1$ and $x=+1$.

At $x=-1$, $y=-3(-1)^2=-3(1)=-3$. The point is $(-1,-3)$.

At $x=1$, $y=-3(1)^2=-3(1)=-3$. The point is $(1,-3)$.

The function has domain $(-\infty,\infty)$ and range $(-\infty,0]$ as seen on the right.

4.031 Graph the function $h(x)=x^2-4$ and state its domain and range.

The function may be written as $y=x^2+0\,x-4$. Comparing it with $y=ax^2+bx+c$, note that $a=1$, $b=0$, $c=-4$.

The x-coordinate of the vertex is $\dfrac{-b}{2a}=\dfrac{-(0)}{2\cdot(1)}=\dfrac{0}{2}=0$.

To find the y-coordinate of the vertex, plug 0 for x into the function:

$y=0^2+0(0)-4=-4$

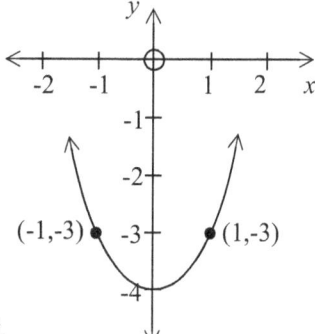

The vertex is at $(0,-4)$ as is the y-intercept. In such cases, find the y-coordinates of two neighboring points, say at $x=-1$ and $x=+1$.

At $x=-1$, $y=(-1)^2-4=1-4=-3$. The point is $(-1,-3)$.

At $x=1$, $y=(1)^2-4=1-4=-3$. The point is $(1,-3)$.

The function has domain $(-\infty,\infty)$ and range $[-4,\infty)$ as seen on the right.

PRACTICE PROBLEMS

Graph the following quadratic functions and mention the domain and range for each:

4.032 $f(x)=-3x^2+6x-1$

4.033 $y=(x-1)(x+7)$

4.034 $j(x)=4x^2$

4.035 $y=(x+4)^2-8$

4.036 $k(x)=25-x^2$

4.037 $y=x-\dfrac{1}{2}x^2$

ABSOLUTE VALUE FUNCTIONS

An **absolute value function** represents a V-shaped graph that resembles either a valley or a peak. Accordingly, at its center lies either the lowermost point or the uppermost point known as the **vertex**. The two straight arms of the V-shape are mirror images of each other on either side of the vertex. Figure on the right shows two sample configurations of an absolute value function.

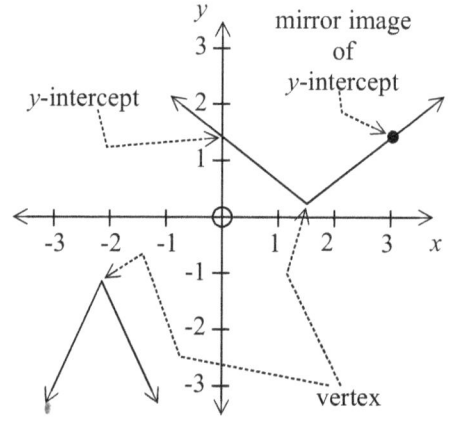

An absolute value function is of the standard form $f(x)=a|x-b|+c$ or $y=a|x-b|+c$ where a is the measure of the steepness of its two arms, and b and c can be used to find the location of the vertex.

To graph an absolute value function, first find the vertex which is at the point (b,c) from the standard form. Then find the y-intercept, whose x-coordinate is 0, and whose y-coordinate can be found by plugging $x=0$ into the function. On the *other side* of the vertex, the mirror image of the y-intercept has its x-coordinate twice that of the vertex, or $2b$, and its y-coordinate same as that of the y-intercept. Draw two rays originating from the vertex, one passing through the y-intercept and the other passing through the mirror image of the y-intercept.

Domain (horizontal reach) of an absolute value function is always $\mathbb{R}$, or $(-\infty,\infty)$, whereas range (vertical reach) is from the y-coordinate of vertex (included) to either positive or negative infinity.

4.038 Graph the absolute value function $f(x)=|x-13|$ and state its domain and range.

The function may be written as $y=1\cdot|x-13|+0$. Comparing it with the standard form $y=a|x-b|+c$ note that $a=1$, $b=13$, $c=0$. The vertex is at (b,c) which is $(13,0)$.

Now find the y-intercept, whose x-coordinate is always 0, and whose y-coordinate can be calculated by plugging 0 for x into the function:

$y=1\cdot|0-13|+0$

$y=|-13|$

$y=13$

So the y-intercept is at $(0,13)$. Its mirror image would be on the other side of the vertex, with x-coordinate twice that of vertex, or 26, and y-coordinate same as that of y-intercept, or 13. The point is $(26,13)$.

Draw rays from the vertex through these two points. The domain is $\mathbb{R}$, or $(-\infty,\infty)$, and the range is $[0,\infty)$.

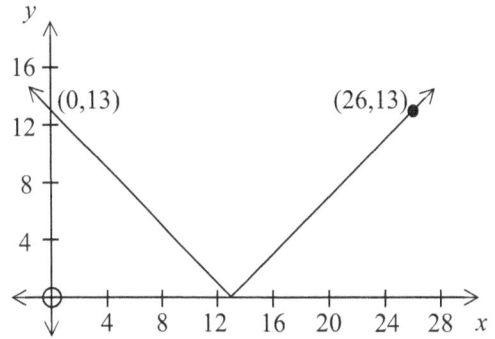

4.039 Graph the function $y=-|x|+5$ and write its domain and range.

The function may be written as $y=-1\cdot|x-0|+5$. Comparing it with the standard form $y=a|x-b|+c$ note that $a=-1$, $b=0$, $c=5$. The vertex is at (b,c) which is $(0,5)$.

Now find the y-intercept, whose x-coordinate is always 0, and whose y-coordinate can be calculated by plugging 0 for x into the function:

$y=-|0|+5$

$y=5$

So the y-intercept is at $(0,5)$, same as vertex. In such cases, find the y-coordinates of two neighboring points, say at $x=-1$ and $x=+1$.

At $x=-1$, $y=-|-1|+5=-1+5=4$. The point is $(-1,4)$.

At $x=1$, $y=-|1|+5=-1+5=4$. The point is $(1,4)$.

Draw rays from the vertex through these two points. The domain is $(-\infty,\infty)$ and range is $(-\infty,5]$. Smaller value first.

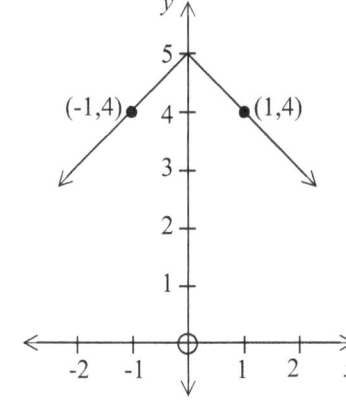

4.040 Graph the function $y=2|x+1|-4$ and mention its domain and range.

Comparing the function with the standard form $y=a|x-b|+c$ note that $a=2$, $b=-1$, $c=-4$. The vertex is at (b,c) which is $(-1,-4)$.

Now find the y-intercept, whose x-coordinate is always 0, and whose y-coordinate can be calculated by plugging 0 for x into the function:

$y=2|0+1|-4$

$y=2|1|-4$

$y=-2$

So the y-intercept is at $(0,-2)$. Its mirror image would be on the other side of the vertex, with x-coordinate twice that of vertex, or -2, and y-coordinate same as that of y-intercept, or -2. The point is $(-2,-2)$.

Draw rays from the vertex through these two points. The domain is $(-\infty,\infty)$ and the range is $[-4,\infty)$.

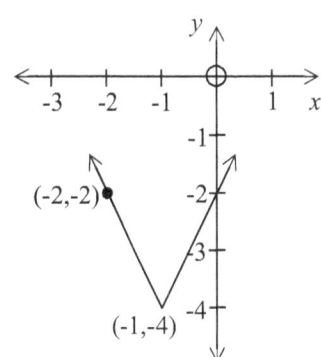

4.041 Graph $y=-3|x|+6$ and write its domain and range.

The function may be written as $y=-3\cdot|x-0|+6$. Comparing it with the standard form $y=a|x-b|+c$ note that $a=-3$, $b=0$, $c=6$. The vertex is at (b,c) which is $(0,6)$.

Now find the y-intercept, whose x-coordinate is always 0, and whose y-coordinate can be calculated by plugging 0 for x into the function:

$y=-3\cdot|0|+6$

$y=6$

So the y-intercept is at $(0,6)$, same as vertex. In such cases, find the y-coordinates of two neighboring points, say at $x=-1$ and $x=+1$.

At $x=-1$, $y=-3|-1|+6=-3+6=3$. The point is $(-1,3)$.

At $x=1$, $y=-3|1|+6=-3+6=3$. The point is $(1,3)$.

Draw rays from the vertex through these two points. The domain is $(-\infty,\infty)$ and range is $(-\infty,6]$. Smaller value first.

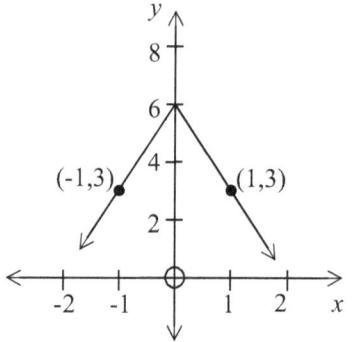

PRACTICE PROBLEMS

Graph the following absolute value functions and state the domain and range for each:

4.042 $y=-7|x|$

4.043 $h(x)=|x-2|$

4.044 $f(x)=-3|x|+10$

4.045 $y=5|x+2|-4$

COMPOSITE FUNCTIONS

Consider the function $f(x)=2x+4$. From what we learned in the section on *evaluating functions* (page 72), we can say that to evaluate $f(x)$ for a specific value of x, just replace x with that value and calculate. For example, to find $f(x)$ at $x=7$, replace x with 7 and calculate. It's written as $f(7)$.

$f(7)=2\cdot(7)+4$

$f(7)=14+4$

$f(7)=18$

By that logic, what is $f(-11)$? That would be:

$f(-11)=2\cdot(-11)+4$

$f(-11)=-22+4$

$f(-11)=-18$

What is $f(a)$? That would be:

$f(a)=2a+4$

What is $f(a^2)$? That would be:

$f(a^2)=2a^2+4$

Now here's another function: $g(x)=x^2$. What if I asked you to find $f(g(x))$? It is read "f of g of x." After taking a close look, you'd replace x with $g(x)$ in the original function $f(x)$. Here's how:

$f(x)=2x+4$

$f(g(x))=2\cdot g(x)+4$

But that's not the final answer, because what exactly is $g(x)$? Well, it's given that $g(x)=x^2$. So put x^2 instead of $g(x)$ in the function $f(g(x))$.

$f(g(x))=2x^2+4$ Final answer.

On similar lines, what is $g(f(x))$? It is read "g of f of x." This time, you'd replace x with $f(x)$ in the original function $g(x)$. Here's how:

$g(x)=x^2$

$g(f(x))=[\ f(x)\ ]^2$

But that's not the final answer, because $f(x)$ happens to be $2x+4$. So put $2x+4$ instead of $f(x)$ in the function $g(f(x))$.

$g(f(x))=(2x+4)^2$

Finally, if you were to calculate $g(f(5))$, how would you proceed? Well, simply look back at the last line; you have already found $g(f(x))$. So just put 5 in the place of x in that function:

$g(f(x))=(2x+4)^2$

$g(f(5))=(2\cdot5+4)^2$

$g(f(5))=(10+4)^2=14^2=196$

The functions $f(g(x))$ and $g(f(x))$ are called **composite functions** because each of them is a *function of a function.*

4.046 Given that $f(x)=x-7$ and $g(x)=2x^3$, find the following composite functions:
$f(18)$, $g(4)$, $f(g(x))$, $g(f(x))$, $g(f(5))$.

To find $f(18)$, put 18 in the place of x in $f(x)$:

$f(x)=x-7$

$f(18)=18-7$

$f(18)=11$

To find $g(4)$, plug 4 for x in $g(x)$:

$g(x)=2x^3$

$g(4)=2\cdot4^3$

$g(4)=2\cdot64=128$

To find $f(g(x))$, put $g(x)$ in the place of x in the original function $f(x)$:

$f(x)=x-7$

$f(g(x))=g(x)-7$

But since $g(x)$ is given as $2x^3$, put that in the above composite function:

$f(g(x))=2x^3-7$ This is the answer.

You might be tempted to put $2x^3$ on the left side also and present your answer as $f(2x^3)=2x^3-7$. While it's not wrong, it's not in the desired format. Remember, the question is asking you to find $f(g(x))$, not $f(2x^3)$.

To find $g(f(x))$, put $f(x)$ in the place of x in the original function $g(x)$:

$g(x)=2x^3$

$g(f(x))=2\cdot[\ f(x)\ ]^3$

But $f(x)$ happens to be $x-7$. So put that in the above composite function:

$g(f(x))=2(x-7)^3$ This is the answer.

To find $g(f(5))$, simply plug 5 into $g(f(x))$:

$g(f(x))=2(x-7)^3$

$g(f(5))=2(5-7)^3$

$g(f(5))=2(-2)^3=2(-8)=-16$

4.047 Given that $h(x)=x+4$ and $f(x)=2^x$, find the following functions:

$h(-6),\ h(a),\ f(0),\ f(h(x)),\ h(f(1))$.

To find $h(-6)$, plug –6 for x in $h(x)$:

$h(x)=x+4$

$h(-6)=-6+4$

$h(-6)=-2$

To find $h(a)$, write a instead of x in $h(x)$:

$h(a)=a+4$

To find $f(0)$, put 0 for x in $f(x)$:

$f(x)=2^x$

$f(0)=2^0=1$ Refer page 43, law 6.

To find $f(h(x))$, put $h(x)$ in the place of x in the original function $f(x)$:

$f(x)=2^x$

$f(h(x))=2^{h(x)}$

But since $h(x)$ is given as $x+4$, put that in the above composite function:

$f(h(x))=2^{x+4}$

$f(h(x))=2^x \cdot 2^4$ Refer page 43, law 1.

$f(h(x))=2^x \cdot 16$

$f(h(x))=16 \cdot (2)^x$ This is the answer.

To find $h(f(1))$, first find $h(f(x))$. Then plug 1 into it. To first find $h(f(x))$, put $f(x)$ in the place of x in the original function $h(x)$:

$h(x)=x+4$

$h(f(x))=f(x)+4$

But since $f(x)$ is given as 2^x, put that in the above composite function:

$h(f(x))=2^x+4$

$h(f(1))=2^1+4$

$h(f(1))=2+4=6$

4.048 If $g(x)=4x^2$ and $h(x)=2x-1$, find $h\left(\frac{1}{2}\right),\ g(b),\ h(g(3)),\ g(h(-1))$.

To find $h\left(\frac{1}{2}\right)$, put $\frac{1}{2}$ in the place of x in $h(x)$:

$h(x)=2x-1$

$h\left(\frac{1}{2}\right)=2 \cdot \frac{1}{2}-1=1-1=0$

To find $g(b)$, write b in the place of x in $g(x)$:

$$g(b)=4b^2$$

To find $h(g(3))$, first find $h(g(x))$. Then plug 3 into it. To first find $h(g(x))$, put $g(x)$ in the place of x in the original function $h(x)$:

$$h(x)=2x-1$$
$$h(g(x))=2\cdot g(x)-1$$

But since $g(x)$ is given as $4x^2$, put that in the above composite function:

$$h(g(x))=2\cdot(4x^2)-1$$
$$h(g(x))=8x^2-1$$
$$h(g(3))=8\cdot3^2-1$$
$$h(g(3))=8\cdot9-1=72-1=71$$

To find $g(h(-1))$, first find $g(h(x))$. Then plug -1 into it. To first find $g(h(x))$, put $h(x)$ in the place of x in the original function $g(x)$:

$$g(x)=4x^2$$
$$g(h(x))=4[\,h(x)\,]^2$$

But $h(x)$ happens to be $2x-1$. So put that in the above composite function:

$$g(h(x))=4(2x-1)^2$$
$$g(h(-1))=4[2(-1)-1]^2$$
$$g(h(-1))=4(-2-1)^2=4(-3)^2=4(9)=36$$

PRACTICE PROBLEMS

4.049 Given that $f(x)=5x$ and $g(x)=x+7$, find $f(-50)$, $g(24)$, $f(g(x))$, $g(f(a))$.

4.050 If $g(x)=x^2-7$ and $h(x)=2x$, find $g(-3)$, $h(-10)$, $g(h(4))$, $h(g(x))$.

4.051 If $f(x)=x+2$ and $h(x)=3^x$, find $f(h(4))$, $h(f(c))$.

INVERSE OF A FUNCTION

Imagine a function $f(x)$, maybe an exponential function, as shown on the right (the upper curve). If you take its mirror image about a 45° inclined line passing through the origin, then you'd get the **inverse** of $f(x)$ shown in the form of the lower curve. It is written as $f^{-1}(x)$ and is read "f inverse of x." Note that the -1 in $f^{-1}(x)$ is not an exponent; it's just the way mathematicians decided to write an inverse of a function. Therefore, there's no such thing as $f^{-2}(x)$, $f^{-3}(x)$ etc.

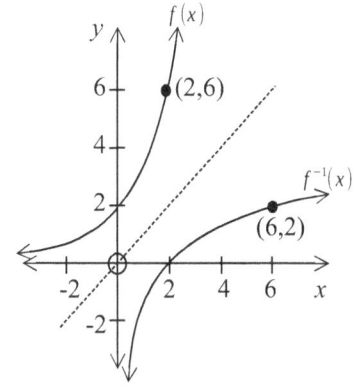

The curve of $f^{-1}(x)$ is obtained when the x and y coordinates of *every single point* on the curve of $f(x)$ are exchanged. For example, the point (2,6) becomes (6,2); the point (0,2) becomes (2,0), and so on.

As you can tell, drawing inverses of functions is quite easy: Simply *invert* them over a 45° line through the origin. In this book, however, we'll learn how to find inverses algebraically.

The procedure to determine the inverse of a function algebraically involves only a few steps: First, make sure the function is written in the form $y=$ not $f(x)=$ Then exchange x and y. Next, solve for y. Finally, replace y with $f^{-1}(x)$. The examples below show how.

4.052 Find the inverse of $f(x)=-2x+8$.

First, write the function as $y=-2x+8$.

Then exchange x and y to get $x=-2y+8$.

Next, solve for y:

$$\underset{-8}{x}=-2y\underset{-8}{+8}$$

$$x-8=-2y$$

$$\frac{x-8}{-2}=\frac{-2y}{-2}$$

$$\frac{x}{-2}-\frac{8}{-2}=y$$

$$-\frac{x}{2}+4=y$$

Finally, replace y with $f^{-1}(x)$.

$$-\frac{x}{2}+4=f^{-1}(x)$$

$$f^{-1}(x)=-\frac{x}{2}+4$$

4.053 Find the inverse of $g(x)=\frac{1}{2}x+5$.

First, write the function as $y=\frac{1}{2}x+5$.

Then exchange x and y to get $x=\frac{1}{2}y+5$.

Next, solve for y:

$$\underset{-5}{x}=\frac{1}{2}y+5\underset{-5}{}$$

$$x-5=\frac{1}{2}y$$

$$\frac{x-5}{\frac{1}{2}}=\frac{\frac{1}{2}y}{\frac{1}{2}}$$

$$(x-5)\cdot\frac{2}{1}=y$$

$$2x-10=y$$

Finally, replace y with $g^{-1}(x)$.

$$2x-10=g^{-1}(x)$$

$$g^{-1}(x)=2x-10$$

4.054 Find the inverse of $h(x)=9x^2-4$.

First, write the function as $y=9x^2-4$.

Now exchange x and y to get $x=9y^2-4$.

Next, solve for y:

$$\underset{+4}{x}=9y^2\underset{+4}{-4}$$

$$x+4=9y^2$$

$$\frac{x+4}{9}=\frac{9y^2}{9}$$

$$\frac{x+4}{9}=y^2$$

$$\sqrt{\frac{x+4}{9}}=\sqrt{y^2}$$

$$\sqrt{\frac{x+4}{9}}=y$$

$$\frac{\sqrt{x+4}}{\sqrt{9}}=y$$

$$\frac{\sqrt{x+4}}{3}=y$$

Finally, replace y with $h^{-1}(x)$.

$$\frac{\sqrt{x+4}}{3}=h^{-1}(x)$$

$$h^{-1}(x)=\frac{\sqrt{x+4}}{3}$$

PRACTICE PROBLEMS

Find inverses of the following functions:

4.055 $p(x)=\frac{1}{4}x^2$

4.056 $f(x)=-6x+7$

4.057 $q(x)=16x^2$

CHAPTER EXERCISES

4.058 If $f(x)=3x^2-8$, then evaluate $f(-2)$, $f(0)$, $f(c)$, $f(3)$.

4.059 Given $f(x)=2x-1$ and $g(x)=3x^2$, determine the following:
$f(10)$, $f(-3)$, $f(g(x))$, $g(f(0))$

4.060 Graph the function $f(x)=2x+5$.

4.061 Graph $h(x)=-\dfrac{x}{4}+6$

4.062 Graph $g(x)=(x+5)(3-x)$ and state its domain and range.

4.063 Graph $p(x)=(x-1)^2+11$ and write its domain and range.

4.064 Graph $f(x)=2\cdot\left(\dfrac{1}{2}\right)^x$ and mention its domain and range.

4.065 Graph $q(x)=-5(2)^x$ and write its domain and range.

4.066 Graph $h(x)=4|x+7|$ and state its domain and range.

4.067 Graph $f(x)=-5|x|+8$ and mention its domain and range.

4.068 Find the inverse of $h(x)=25x^2$.

4.069 Find $f^{-1}(x)$ if $f(x)=\dfrac{1}{3}x+1$.

CHAPTER 5
INEQUALITIES

This chapter builds on top of Chapter 3 which was about equations. Unlike an equation, an **inequality** has its two sides *not equal* to each other. One side may be less than (<), less than or equal to (≤), greater than (>), or greater than or equal to (≥) the other side.

Consider the fact that 4 is greater than 3. Mathematically, we can write it as $4 > 3$. It is also correct to write it as $4 \geq 3$, because while 4 is not equal to 3, it is still greater than 3. So, the statement is valid. Likewise, $17 \leq 20$; $-6 < -1$; $35 \geq -9$ are all valid inequality statements.

Inequalities are solved just like equations, except that the inequality sign reverses when both sides of the inequality are multiplied or divided by a *negative number*. For example, we know that $-2 < 4$. Multiplying both sides by a negative number, say -1, we get $2 > -4$. Notice the inequality sign reversed.

An inequality can have any number of variables. In this book, however, we will study those containing only one or two variables.

INEQUALITIES IN ONE VARIABLE

Back when we solved equations in one variable, we got specific values for it such as $x=4$, $x=-9$ etc. However, when inequalities in one variable are solved, we get an *interval* of values. For example, $x \geq 5$ means x can be 5 or greater, all the way to infinity. That would be the interval $[5,\infty)$. Similarly, $x < 1$ means x can be less than 1, all the way to negative infinity. In interval notation, that would be $(-\infty,1)$. Smaller value first.

The solution of an inequality in one variable can be shown on the number line. For example, $x \geq 5$ or the interval $[5,\infty)$ can be graphed as:

Likewise, $x < 1$ or the interval $(-\infty,1)$ can be graphed as:

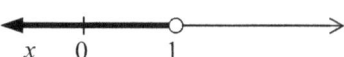

At the boundary of an interval, a solid circle is shown when the value is *included* (like at 5). A hollow circle is shown when the value is *not included* (like at 1). The interval should be darkened. Infinity or negative infinity are never shown on the graph; instead, an arrow suffices. It's also a good practice to show 0 for reference when only one other number appears at the boundary of the solution.

5.001 $3x+8\geq29$

Solving along the lines of an equation, we'd do:

$3x+8\geq29$
$\underset{-8}{}\underset{-8}{}$

$3x\geq21$

$\dfrac{3x}{3}\geq\dfrac{21}{3}$

$x\geq7$ which is the interval $[7,\infty)$.

It is graphed as shown below. The solid circle at 7 means 7 is *included*. If 7 weren't included, we'd use a hollow circle.

5.002 $2-6x<14$

Just like in an equation, isolate the x step by step.

$2-6x<14$
$\underset{-2}{}\phantom{-6x<}\underset{-2}{}$

$-6x<12$

$\dfrac{-6x}{-6}>\dfrac{12}{-6}$

The inequality sign reversed because we divided by a negative number.

$x>-2$ which is the interval $(-2,\infty)$.

The graph looks as shown below. Notice the hollow circle at -2 because -2 is *not included*.

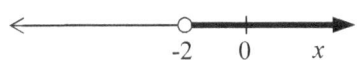

5.003 $-5\leq19-2x$

It's always a good idea to get the x term on the left side.

$-5\leq19-2x$
$\underset{+2x}{}\underset{+2x}{}$

$-5+2x\leq19$
$\underset{+5}{}\underset{+5}{}$

$2x\leq24$

$\dfrac{2x}{2}\leq\dfrac{24}{2}$

$x\leq12$ which is the interval $(-\infty,12]$.
The graph is shown below.

5.004 $x+8>32+13x$

Get the x terms on the left, and the constants on the right.

$x+8>32+13x$
$\underset{-13x}{}\underset{-13x}{}$

$-12x+8>32$
$\underset{-8}{}\underset{-8}{}$

$-12x>24$

$\dfrac{-12x}{-12}<\dfrac{24}{-12}$

Division by a negative number reversed the inequality sign.

$x<-2$ which is the interval $(-\infty,-2)$.

The graph is shown below.

5.005 $2(3x+1)<2-(1-6x)$

Distribute and simplify first. Be mindful of the negative signs.

$6x+2<2-1+6x$

$6x+2<1+6x$

$6x+2<1+6x$
$\underset{-6x}{}\underset{-6x}{}$

$2<1$

2 is less than 1? Really? That's *never* true. It means the inequality has no solution, and therefore there's nothing to graph. Don't be surprised if you come across such an example.

5.006 $7(3+x)+6\geq4x+3(9+x)$

Distribute and simplify.

$21+7x+6\geq4x+27+3x$

$27+7x\geq7x+27$

$27+7x\geq7x+27$
$\underset{-7x}{}\underset{-7x}{}$

$27\geq27$

Since 27 is equal to 27, this inequality is considered *true* regardless of the value of x. (It would be considered *false* if it were $27>27$.) So x can take on just about any real number value. The solution is $\mathbb{R}$, or $(-\infty,\infty)$. The graph would be the entire number line as shown below.

5.007 $5x-(5-x)>4x+4(x-1)-1$

Distribute, simplify and solve.

$5x-5+x>4x+4x-4-1$

$6x-5>8x-5$

$\begin{array}{cc}6x-5>8x-5\\-8x \quad\quad -8x\end{array}$

$-2x-5>-5$

$\begin{array}{cc}-2x-5>-5\\+5 \quad +5\end{array}$

$-2x>0$

$\dfrac{-2x}{-2}<\dfrac{0}{-2}$

Division by a negative number reversed the inequality sign.

$x<0$ which is the interval $(-\infty,0)$.

The graph is shown below.

PRACTICE PROBLEMS

5.008 $4x+13\geq6$

5.009 $-x+7\geq5x+1$

5.010 $-2x+10\leq8-4x$

5.011 $x-3(x-3)>2x+5$

5.012 $-2(5-3x)+1>18x+3(1-4x)$

5.013 $x+19-3(x+1)<2(x-8)$

5.014 $x-2\big(x-2(x-2)\big)<3(x-2)$

COMPOUND INEQUALITIES

A **compound inequality** results when two or more inequalities are considered together. In this book, we will focus on only two inequalities at a time when writing a compound inequality. For example, suppose $x>5$ and $x\leq26$. How do you represent them together on the same graph? Well, $x>5$ also means that $5<x$. So, $5<x$ and $x\leq26$ together would be $5<x\leq26$. Meaning, x can be within the range 5 to 26, where 5 is *not* included, and 26 *is* included. The graph would be as shown below. Notice the hollow circle at 5 and the solid circle at 26.

The interval $(5,26]$ is called a **continuous interval** because it is included *between* two numbers, not on the *outside* of the two numbers.

Consider another instance in which $x<-8$ or $x>14$. It is not possible to combine them into one statement. However, together they do represent a compound inequality. Its graph is shown below.

The intervals $(-\infty,-8)$ and $(14,\infty)$ together comprise its solution set, and are written in a single statement as $(-\infty,-8)$ U $(14,\infty)$, where U represents *union* of the two intervals. Alternatively, you may write the statement as $x<-8$ OR $x>14$. The word "OR" is used because any specific value of x you pick can either be less than -8 *or* more than 14. It can't be both at the same time, can it? This type of interval is called a **discontinuous interval** because the values of x are on the *outside* of the two numbers, not between the two numbers.

Now let's solve a few compound inequalities and draw their graphs.

5.015 $-12 < 3x - 6 < 9$

Solve it like a regular inequality, but keep in mind that whatever you do to one side, you must do to *all three sides* of it. Isolate x step by step.

$$\begin{array}{ccc} -12 < 3x - 6 < 9 \\ {\scriptstyle +6} \quad {\scriptstyle +6} \quad {\scriptstyle +6} \end{array}$$

$$-6 < 3x < 15$$

$$\frac{-6}{3} < \frac{3x}{3} < \frac{15}{3}$$

$-2 < x < 5$ which is the interval $(-2, 5)$.

The graph is shown below.

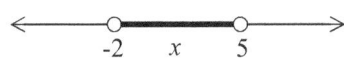

5.016 $-7 \le 1 - 2x < 5$

Isolate x step by step.

$$\begin{array}{ccc} -7 \le 1 - 2x < 5 \\ {\scriptstyle -1} \quad {\scriptstyle -1} \quad {\scriptstyle -1} \end{array}$$

$$-8 \le -2x < 4$$

$$\frac{-8}{-2} \ge \frac{-2x}{-2} > \frac{4}{-2}$$

Division by a negative number reversed the inequality signs.

$$4 \ge x > -2$$

The inequality appears to show values in decreasing order. They need to be in *increasing* order for us to correctly graph them on the number line. So, reverse the values *and* inequality signs, and rewrite.

$-2 < x \le 4$ which is the interval $(-2, 4]$.

The graph is shown below.

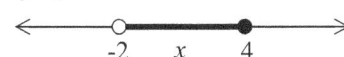

5.017 $-10 \le \dfrac{x+1}{4} \le 20$

Isolate x step by step.

$$(-10) \cdot 4 \le \frac{x+1}{4} \cdot 4 \le 20 \cdot 4$$

$$-40 \le x + 1 \le 80$$

$$\begin{array}{ccc} -40 \le x + 1 \le 80 \\ {\scriptstyle -1} \quad {\scriptstyle -1} \quad {\scriptstyle -1} \end{array}$$

$-41 \le x \le 79$ which is $[-41, 79]$.

The graph is shown below.

5.018 $5x + 10 < 25$ OR $2x - 1 > 13$

Solve the inequalities separately as you would, with the word "OR" between them.

$$\begin{array}{ccc} 5x + 10 < 25 & \text{OR} & 2x - 1 > 13 \\ {\scriptstyle -10} \quad {\scriptstyle -10} & & {\scriptstyle +1} \quad {\scriptstyle +1} \end{array}$$

$$5x < 15 \qquad \text{OR} \qquad 2x > 14$$

$$\frac{5x}{5} < \frac{15}{5} \qquad \text{OR} \qquad \frac{2x}{2} > \frac{14}{2}$$

$$x < 3 \qquad \text{OR} \qquad x > 7$$

The interval is $(-\infty, 3) \cup (7, \infty)$.

The graph is shown below.

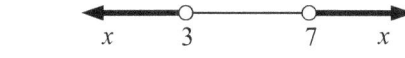

5.019 $-2x - 17 \ge 7$ OR $\dfrac{x-3}{15} > 1$

Isolate x in both inequalities, step by step.

$$\begin{array}{ccc} -2x - 17 \ge 7 & \text{OR} & \frac{x-3}{15} \cdot 15 > 1 \cdot 15 \\ {\scriptstyle +17} \quad {\scriptstyle +17} & & \end{array}$$

$$-2x \ge 24 \qquad \text{OR} \qquad x - 3 > 15$$

$$\frac{-2x}{-2} \le \frac{24}{-2} \qquad \text{OR} \qquad \begin{array}{c} x - 3 > 15 \\ {\scriptstyle +3} \quad {\scriptstyle +3} \end{array}$$

Division by a negative number reversed the first inequality sign.

$$x \le -12 \qquad \text{OR} \qquad x > 18$$

The interval is $(-\infty, -12] \cup (18, \infty)$.

The graph is shown below.

PRACTICE PROBLEMS

5.020 $-2 \le 1 - 3x < 2$

5.021 $x + 3 \ge 4$ OR $8 - x > 10$

5.022 $0 < 2(x + 11) \le 50$

5.023 $\dfrac{2 + 3x}{4} \le 5$ OR $5x + 5 \ge 50$

5.024 $2 < \dfrac{3x - 12}{6} < 3$

ABSOLUTE VALUE INEQUALITIES

Absolute value inequalities are solved by writing two separate inequalities along the lines of absolute value equations (from page 64). The solutions to the two separate inequalities when put together yield either a *continuous* or a *discontinuous* interval, with the graph being similar to that of a compound inequality.

Please remember that if the modulus (the absolute value) in the inequality turns out to be negative, then the inequality has no solution. That's because absolute value can be positive or zero, but *never* negative.

5.025 $\quad |x+3| < 13$

Write the two separate inequalities:

$$x+3 < 13 \quad \bigg| \quad -(x+3) < 13$$

$$\begin{array}{c} x+3 < 13 \\ {\scriptstyle -3 \quad -3} \\ \hline x < 10 \end{array} \quad \bigg| \quad \begin{array}{c} -x-3 < 13 \\ -x-3 < 13 \\ {\scriptstyle +3 \quad +3} \\ \hline -x < 16 \\ \dfrac{-x}{-1} > \dfrac{16}{-1} \\ x > -16 \end{array}$$

The graph is shown below.
The interval is $(-16, 10)$.

5.026 $\quad |x-8| + 9 \geq 21$

First isolate the modulus (the absolute value) to check if the inequality is valid.

$$\begin{array}{c} |x-8| + 9 \geq 21 \\ {\scriptstyle -9 \quad -9} \\ \hline |x-8| \geq 12 \end{array}$$

The modulus is positive. The inequality is valid. Write the two separate inequalities:

$$x-8 \geq 12 \quad \bigg| \quad -(x-8) \geq 12$$

$$\begin{array}{c} x-8 \geq 12 \\ {\scriptstyle +8 \quad +8} \\ \hline x \geq 20 \end{array} \quad \bigg| \quad \begin{array}{c} -x+8 \geq 12 \\ -x+8 \geq 12 \\ {\scriptstyle -8 \quad -8} \\ \hline -x \geq 4 \\ \dfrac{-x}{-1} \leq \dfrac{4}{-1} \\ x \leq -4 \end{array}$$

The graph is shown below.
The interval is $(-\infty, -4] \cup [20, \infty)$.

5.027 $\quad |2x+5| + 7 \leq 4$

First isolate the modulus.

$$\begin{array}{c} |2x+5| + 7 \leq 4 \\ {\scriptstyle -7 \quad -7} \\ \hline |2x+5| \leq -3 \end{array}$$

The modulus is negative. The inequality is invalid, and has no solution.

5.028 $\quad |2x-11| - 42 > 1$

First isolate the modulus.

$$\begin{array}{c} |2x-11| - 42 > 1 \\ {\scriptstyle +42 \quad +42} \\ \hline |2x-11| > 43 \end{array}$$

The modulus is positive. The inequality is valid. Write the two separate inequalities:

$$2x-11 > 43 \quad \bigg| \quad -(2x-11) > 43$$

$$\begin{array}{c} 2x-11 > 43 \\ {\scriptstyle +11 \quad +11} \\ \hline 2x > 54 \\ \dfrac{2x}{2} > \dfrac{54}{2} \\ x > 27 \end{array} \quad \bigg| \quad \begin{array}{c} -2x+11 > 43 \\ -2x+11 > 43 \\ {\scriptstyle -11 \quad -11} \\ \hline -2x > 32 \\ \dfrac{-2x}{-2} < \dfrac{32}{-2} \\ x < -16 \end{array}$$

The graph is shown below.
The interval is $(-\infty, -16) \cup (27, \infty)$.

PRACTICE PROBLEMS

5.029 $\quad |5x-13| + 1 < 18$

5.030 $\quad |4-3x| + 8 < 6$

5.031 $\quad 7 + |3x-1| \geq 21$

5.032 $\quad |x+6| \leq 9$

INEQUALITIES IN TWO VARIABLES (LINEAR INEQUALITIES)

An inequality in two variables (x and y) with both variables in their first degree is a **linear inequality**. Before graphing such an inequality, solve it for y. The graph will be in the form of a region shaded either above or below a line, the line being either dashed ($-----$) or solid (———).

If a linear inequality looks like either $y > ax + b$ or $y \geq ax + b$, then the region *above* the line would be shaded. If a linear inequality looks like either $y < ax + b$ or $y \leq ax + b$, then the region *below* the line would be shaded.

The line would be drawn *dashed* if the inequality signs are non-inclusive (either $>$ or $<$). The line would be drawn *solid* if the inequality signs are inclusive (either $\geq$ or $\leq$).

Visit page 75 if you need a refresher on how to graph a line given a linear equation or a linear function.

5.033 $\quad y \geq \dfrac{1}{3}x - 2$

In the slope of the line, the run is 3. So go 3 paces to the left and right of 0 on the x-axis. You'll find the coordinates -3 and $+3$. Calculate the y-coordinates at $x = -3$, $+3$, and 0 thinking of the inequality as an *equation*. Plot those points and draw a *solid* line through them. Shade the region *above* it. The table and graph are shown below.

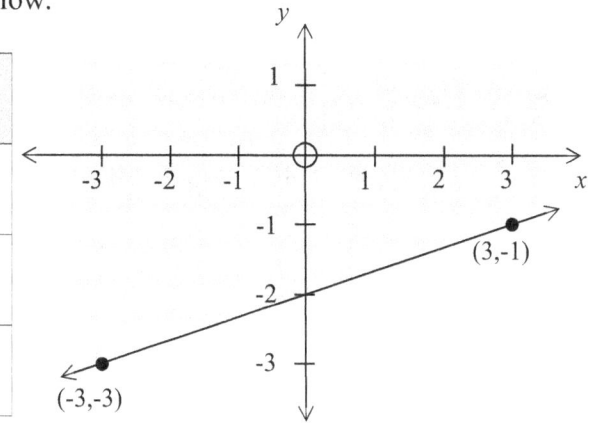

x	$y = \dfrac{1}{3}x - 2$
-3	$y = \dfrac{1}{3} \cdot (-3) - 2 = \dfrac{-3}{3} - 2 = -1 - 2 = -3$
0	$y = \dfrac{1}{3} \cdot (0) - 2 = 0 - 2 = -2$
3	$y = \dfrac{1}{3} \cdot (3) - 2 = \dfrac{3}{3} - 2 = 1 - 2 = -1$

5.034 $\quad y < 2x + 1$

The given inequality can be written as $y < \dfrac{2}{1}x + 1$, in the slope of which, the run is 1. So go 1 step to the left and right of 0 on the x-axis. You'll find the coordinates -1 and $+1$. Calculate the y-coordinates at $x = -1$, $+1$, and 0 thinking of the given inequality as an *equation*. Plot those points and draw a *dashed* line through them. Shade the region *below* it. The table and graph are shown below.

x	$y = 2x + 1$
-1	$y = 2 \cdot (-1) + 1 = -2 + 1 = -1$
0	$y = 2 \cdot (0) + 1 = 0 + 1 = 1$
1	$y = 2 \cdot (1) + 1 = 2 + 1 = 3$

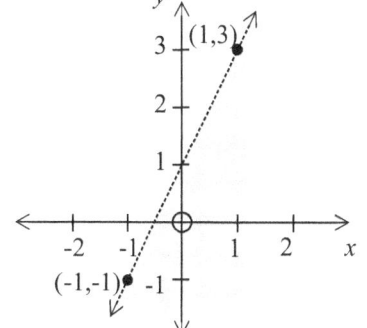

5.035 $2x+3y \le 12$

First solve the inequality for y.

$$2x+3y \le 12$$
$$_{-2x} \qquad _{-2x}$$
$$3y \le 12-2x$$
$$3y \le -2x+12$$

$$\longrightarrow \qquad \frac{3y}{3} \le \frac{-2x+12}{3}$$

$$y \le -\frac{2}{3}x+4$$

In the slope of the line, the run is 3. So go 3 paces to the left and right of 0 on the x-axis. You'll find the coordinates –3 and +3. Calculate the y-coordinates at $x=-3$, $+3$, and 0 thinking of the inequality as an *equation*. Plot those points and draw a *solid* line through them. Shade the region *below* it. The table and graph are shown below.

x	$y=-\dfrac{2}{3}x+4$
-3	$y=-\dfrac{2}{3}\cdot(-3)+4=\dfrac{6}{3}+4=2+4=6$
0	$y=-\dfrac{2}{3}\cdot(0)+4=0+4=4$
3	$y=-\dfrac{2}{3}\cdot(3)+4=-\dfrac{6}{3}+4=-2+4=2$

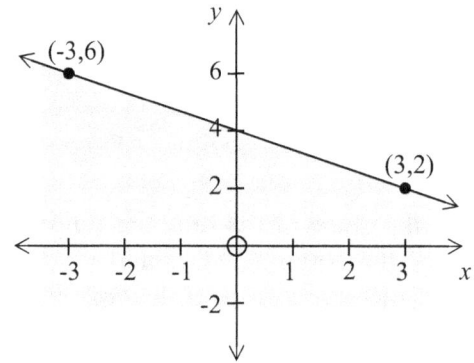

5.036 $\dfrac{1}{2}x-y<1$

First solve the inequality for y.

$$\frac{1}{2}x-y<1$$
$$_{-\frac{1}{2}x} \qquad _{-\frac{1}{2}x}$$
$$-y<1-\frac{1}{2}x$$

$$\longrightarrow \qquad -y<-\frac{1}{2}x+1$$

$$\frac{-y}{-1}<\frac{-\dfrac{1}{2}x+1}{-1}$$

$$y>\frac{1}{2}x-1$$

In the slope of the line, the run is 2. So go 2 paces to the left and right of 0 on the x-axis. You'll find the coordinates –2 and +2. Calculate the y-coordinates at $x=-2$, $+2$, and 0 thinking of the inequality as an *equation*. Plot those points and draw a *dashed* line through them. Shade the region *above* it. The table and graph are shown below.

x	$y=\dfrac{1}{2}x-1$
-2	$y=\dfrac{1}{2}\cdot(-2)-1=\dfrac{-2}{2}-1=-1-1=-2$
0	$y=\dfrac{1}{2}\cdot(0)-1=0-1=-1$
2	$y=\dfrac{1}{2}\cdot(2)-1=\dfrac{2}{2}-1=1-1=0$

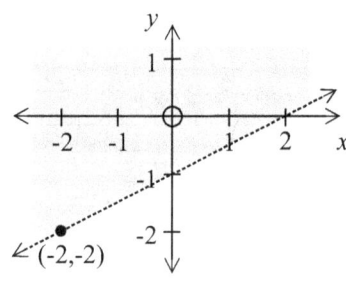

PRACTICE PROBLEMS

5.037 $y \geq \frac{1}{4}x$

5.038 $4x - 2y < 2$

5.039 $y \leq -\frac{3}{4}x + 2$

5.040 $y - x < 3$

CHAPTER EXERCISES

5.041 $14 - 3x \geq 20$

5.042 $x + 5(x-1) \leq 3(2x-1)$

5.043 $2x + 8 > 2(x+4)$

5.044 $-6 \leq \frac{x+5}{3} < 6$

5.045 $\frac{x-4}{9} \leq 1$ OR $2x - 14 \geq 26$

5.046 $5|4x-6| + 13 < 8$

5.047 $|9 - 3x| < 6$

5.048 $2|x+7| - 1 \leq 9$

5.049 $y > -\frac{1}{2}x + 2$

5.050 $3x + y \leq 1$

CHAPTER 6
SEQUENCES AND SERIES

A numeric **sequence** is a progression of numbers in a specific format. It is simply a *list* of numbers. Sequences can be of several types: arithmetic, geometric, quadratic, harmonic etc. In this book, we will only study arithmetic and geometric sequences. When the numbers in a particular sequence are added up, what you get is a **series**. Meaning, a series is the *sum* of that sequence; it is not the sequence itself. So, just to be clear: sequence = list   series = sum of that list

ARITHMETIC SEQUENCE

Consider the sequence 1, 7, 13, 19, ...   Some things you may notice are that it starts with 1, and each term after that is 6 more than the previous term. The sequence doesn't seem to end. Such a sequence of numbers increasing (or decreasing) at a constant pace is called an **arithmetic sequence**. Think of it as a stairway with a uniform step height of 6 inches. After every step, you'd be 6 inches higher (or lower). Since the stairway would never end, think of it as a stairway to heaven (or hell)! An unending sequence is called an **infinite** sequence. A sequence with a limited number of terms would be a **finite** sequence.

In mathematical notation, we use a_1, a_2, a_3, ... , a_n to denote the 1st term, 2nd term, 3rd term, ... , nth term. In the above arithmetic sequence, $a_1 = 1$; $a_2 = 7$; $a_3 = 13$; $a_4 = 19$ and so on. Please note that the notation is not a^1, a^2, a^3, ... because a^1, a^2, a^3, ... would mean exponents of a, not terms of a sequence.

THE EXPLICIT FORMULA

Following the pattern in the above sequence, if I asked you to find the 5th term (a_5), you'd happily tell me it's just 6 more than 19, which is 25. If I asked you for the 13th term (a_{13}), you'd begrudgingly work it out and tell me it's 73. If I asked you to find the 55th term (a_{55}), you'd probably want to spank me. Or maybe not. The point is, there's a formula to find just about any term (the nth term, a_n) in an arithmetic sequence. It's called the **explicit formula**, which is:

$$a_n = a_1 + d(n-1)$$ where a_n is the nth term

a_1 is the 1st term of the sequence

d is the step size, known as the **common difference**

n is the rank of the term

The common difference (d) can be calculated as any term minus the term before it, as in $a_2 - a_1$ or $a_3 - a_2$ or $a_4 - a_3$ etc.

6.001 In the sequence 2, 5, 8, 11, … find a_9 and a_{30}.

Let's use the explicit formula to find each of them, the step size (common difference, d) being 3.

To find the 9th term (a_9):

$a_n = a_1 + d(n-1)$

$a_9 = 2 + 3(9-1)$

$a_9 = 2 + 3(8)$

$a_9 = 26$

To find the 30th term (a_{30}):

$a_n = a_1 + d(n-1)$

$a_{30} = 2 + 3(30-1)$

$a_{30} = 2 + 3(29)$

$a_{30} = 89$

6.002 In the sequence −7, −5, −3, −1, … find a_6 and a_{16}.

The step size (common difference, d) is $a_2 - a_1 = -5 - (-7) = -5 + 7 = 2$

To find the 6th term (a_6):

$a_n = a_1 + d(n-1)$

$a_6 = -7 + 2(6-1)$

$a_6 = -7 + 2(5)$

$a_6 = 3$

To find the 16th term (a_{16}):

$a_n = a_1 + d(n-1)$

$a_{16} = -7 + 2(16-1)$

$a_{16} = -7 + 2(15)$

$a_{16} = 23$

6.003 In the sequence 15, 11, 7, 3, … find a_{10} and a_{18}.

Notice that the terms are decreasing in value, the common difference (d) being −4, not 4.

To find the 10th term (a_{10}):

$a_n = a_1 + d(n-1)$

$a_{10} = 15 + (-4)(10-1)$

$a_{10} = 15 - 4(9)$

$a_{10} = -21$

To find the 18th term (a_{18}):

$a_n = a_1 + d(n-1)$

$a_{18} = 15 + (-4)(18-1)$

$a_{18} = 15 - 4(17)$

$a_{18} = -53$

6.004 In the sequence −1.75, −2, −2.25, −2.5, … find a_{13} and a_{20}.

The common difference (d) is $a_2 - a_1 = -2 - (-1.75) = -2 + 1.75 = -0.25$

To find the 13th term (a_{13}):

$a_n = a_1 + d(n-1)$

$a_{13} = -1.75 + (-0.25)(13-1)$

$a_{13} = -1.75 - 0.25(12)$

$a_{13} = -4.75$

To find the 20th term (a_{20}):

$a_n = a_1 + d(n-1)$

$a_{20} = -1.75 + (-0.25)(20-1)$

$a_{20} = -1.75 - 0.25(19)$

$a_{20} = -6.5$

PRACTICE PROBLEMS

In each of the following sequences, find the indicated terms:

6.005 22, 19, 16, 13, … Find a_{15} and a_{50}.

6.006 −9, −5, −1, 3, … Find a_8 and a_{23}.

6.007 1.4, 2.7, 4, 5.3, … Find a_9 and a_{13}.

6.008 64, 57, 50, 43, … Find a_{25} and a_{33}.

THE RECURSIVE FORMULA

A recursive formula tells you how to obtain the next term in a sequence. In other words, it tells you how to obtain the nth term (a_n) from its predecessor, the $(n-1)^{th}$ term. To understand how to write a recursive formula, let's take up the following sequence from earlier in this chapter: 1, 7, 13, 19, …

Each term is increasing by 6 over its previous term. The common difference is 6. Meaning, to find any term add 6 to the previous term. Likewise, to find a_n add 6 to a_{n-1}, the recursive formula being $a_n = a_{n-1} + 6$. That's all there is to it!

Please note that it is also correct to write the recursive formula as $a_{n+1} = a_n + 6$. Meaning, to obtain the $(n+1)^{th}$ term add 6 to the nth term. But let's consistently stick to writing $a_n = \ldots$ instead of $a_{n+1} = \ldots$

6.009 Find the recursive formula for the sequence 4, 9, 14, 19, …

Observe that each term is 5 greater than the previous term, the common difference being 5. To obtain the nth term (a_n) add 5 to the $(n-1)^{th}$ term (a_{n-1}). The recursive formula is: $a_n = a_{n-1} + 5$

6.010 Write the recursive formula for the sequence 2.1, 2.5, 2.9, 3.3, …

The common difference is: $d = a_2 - a_1 = 2.5 - 2.1 = 0.4$ To obtain the nth term (a_n) add 0.4 to the $(n-1)^{th}$ term (a_{n-1}). The recursive formula is: $a_n = a_{n-1} + 0.4$

6.011 Find the recursive formula for the sequence –3, –5, –7, –9, …

The common difference is: $d = a_2 - a_1 = -5 - (-3) = -5 + 3 = -2$ To obtain the nth term (a_n) add –2 to the $(n-1)^{th}$ term (a_{n-1}). The recursive formula is: $a_n = a_{n-1} + (-2)$ or simply $a_n = a_{n-1} - 2$

PRACTICE PROBLEMS

6.012 Write recursive formulas for examples 6.001 through 6.004.

6.013 Find the recursive formula for the sequence 68, 72, 76, 80, …

6.014 Write the recursive formula for the sequence 1.2, 0.7, 0.2, –0.3, …

ARITHMETIC SERIES

As we discussed earlier in this chapter, a sequence is a *list* of numbers, whereas a series is the *sum* of that list. Consider the arithmetic sequence 4, 7, 10, 13, … The sum of its terms would be the infinite **arithmetic series** 4+7+10+13+… If you try to add up all its terms, you'd head toward infinity (∞), an undefined value. Thus, it's not possible to evaluate an infinite arithmetic series. But we can find the sum of a finite portion of it, maybe the first 5 terms, or the first 30 terms etc. In mathematical notation, the sum of the first n terms in a series is written as S_n. The formula to evaluate an arithmetic series is:

$$S_n = \left(\frac{a_1 + a_n}{2} \right) \cdot n \qquad \text{where}$$

S_n is the sum of the first n terms
a_1 is the first term
a_n is the nth term
n is the number of terms to be added up

Using this formula, let's evaluate (on the next page) S_{10} for the above series.

To calculate S_{10} in the series $4+7+10+13+...$, we see that a_1 is 4, n is 10, but a_n (which is a_{10}) is not known. So let's first find a_{10} using the formula for the nth term of an arithmetic sequence from page 100.

$a_n = a_1 + d(n-1)$ where the common difference (d) is 3

$a_{10} = 4 + 3(10-1)$

$a_{10} = 4 + 3(9)$

$a_{10} = 31$

We can now use the formula for sum.

$$S_n = \left(\frac{a_1 + a_n}{2}\right) \cdot n$$

$$S_{10} = \left(\frac{a_1 + a_{10}}{2}\right) \cdot 10$$

$$S_{10} = \left(\frac{4 + 31}{2}\right) \cdot 10$$

$$S_{10} = \frac{35}{2} \cdot 10 = 175$$

In these problems on finding the sum, follow the order d, a_n and S_n. daS, the opposite of Sad.

6.015 Find the sum of the first 15 terms (S_{15}) of the series $(-2)+(-5)+(-8)+(-11)+...$

In the given series, common difference is:
$d = a_2 - a_1 = -5 - (-2) = -5 + 2 = -3$

Now let's find a_n (which is a_{15}). We'll need it to calculate S_{15}.

$a_n = a_1 + d(n-1)$

$a_{15} = -2 + (-3)(15-1)$

$a_{15} = -2 - 3(14)$

$a_{15} = -44$

Finally, use the formula for the sum S_n.

$$S_n = \left(\frac{a_1 + a_n}{2}\right) \cdot n$$

$$S_{15} = \left(\frac{a_1 + a_{15}}{2}\right) \cdot 15$$

$$S_{15} = \left(\frac{-2 + (-44)}{2}\right) \cdot 15$$

$$S_{15} = \left(\frac{-46}{2}\right) \cdot 15 = -345$$

6.016 Find S_{24} for the series $3+10+17+24+...$

Let's follow the order d, a_n, S_n.

$d = a_2 - a_1 = 10 - 3 = 7$

$a_n = a_1 + d(n-1)$

$a_{24} = 3 + 7(24-1)$

$a_{24} = 3 + 7(23)$

$a_{24} = 164$

$$S_n = \left(\frac{a_1 + a_n}{2}\right) \cdot n$$

$$S_{24} = \left(\frac{a_1 + a_{24}}{2}\right) \cdot 24$$

$$S_{24} = \left(\frac{3 + 164}{2}\right) \cdot 24$$

$$S_{24} = \left(\frac{167}{2}\right) \cdot 24 = 2004$$

6.017 Evaluate S_{12} for the following sequence:
1.7, 3.3, 4.9, 6.5, …

Let's go in the order d, a_n, S_n.

$d = a_2 - a_1 = 3.3 - 1.7 = 1.6$

$a_n = a_1 + d(n-1)$

$a_{12} = 1.7 + 1.6(12-1)$

$a_{12} = 1.7 + 1.6(11)$

$a_{12} = 19.3$

$S_n = \left(\dfrac{a_1 + a_n}{2}\right) \cdot n$

$S_{12} = \left(\dfrac{a_1 + a_{12}}{2}\right) \cdot 12$

$S_{12} = \left(\dfrac{1.7 + 19.3}{2}\right) \cdot 12$

$S_{12} = \left(\dfrac{21}{2}\right) \cdot 12 = 126$

6.018 Evaluate S_{30} for the following sequence:
65, 63.5, 62, 60.5, …

Let's solve in the order d, a_n, S_n.

$d = a_2 - a_1 = 63.5 - 65 = -1.5$

$a_n = a_1 + d(n-1)$

$a_{30} = 65 + (-1.5)(30-1)$

$a_{30} = 65 - 1.5(29)$

$a_{30} = 21.5$

$S_n = \left(\dfrac{a_1 + a_n}{2}\right) \cdot n$

$S_{30} = \left(\dfrac{a_1 + a_{30}}{2}\right) \cdot 30$

$S_{30} = \left(\dfrac{65 + 21.5}{2}\right) \cdot 30$

$S_{30} = \left(\dfrac{86.5}{2}\right) \cdot 30 = 1297.5$

PRACTICE PROBLEMS

6.019 Calculate S_{11} for the sequence 2, 6, 10, 14, …

6.020 Evaluate S_{20} for the series $4.6 + 4.1 + 3.6 + 3.1 + …$

6.021 Find S_{17} for the sequence –7.3, –7.1, –6.9, –6.7, …

6.022 Evaluate S_8 for the series $(-32) + (-34) + (-36) + (-38) + …$

GEOMETRIC SEQUENCE

Consider the sequence 2, 6, 18, 54, … Notice that the numbers are not equally spaced out like those in an arithmetic sequence; instead, each term is being multiplied by 3 to obtain the next term. Such a multiplier (the 3) is called the **common ratio**, and such a sequence is called a **geometric sequence**.

EXPLICIT AND RECURSIVE FORMULAS

The **explicit formula** to determine the nth term (a_n) of a geometric sequence is:

$a_n = a_1 \cdot r^{n-1}$ where a_n is the nth term
a_1 is the 1st term of the sequence
r is the common ratio (the multiplier)
n is the rank of the term

The common ratio (r) can be calculated by dividing any term by the term before it, as in $a_2 \div a_1$ or $a_3 \div a_2$ or $a_4 \div a_3$ etc.

The **recursive formula** shows how to obtain the nth term (a_n) from its predecessor, the $(n-1)^{th}$ term (a_{n-1}). In the case of the sequence 2, 6, 18, 54, … a_n can be obtained by multiplying a_{n-1} by 3. So the recursive formula would be: $a_n = a_{n-1} \times 3$

6.023 For the sequence 3, 4.5, 6.75, 10.125, … write the recursive formula. Use the explicit formula to find a_6 and a_8.

The multiplier (common ratio, r) is $\dfrac{a_2}{a_1} = \dfrac{4.5}{3} = 1.5$ Recursive formula: $a_n = a_{n-1} \times 1.5$

To find the 6th term (a_6):

$a_n = a_1 \cdot r^{n-1}$

$a_6 = 3 \cdot (1.5)^{6-1}$

$a_6 = 3 \cdot (1.5)^5 = 22.78125$

To find the 8th term (a_8):

$a_n = a_1 \cdot r^{n-1}$

$a_8 = 3 \cdot (1.5)^{8-1}$

$a_8 = 3 \cdot (1.5)^7 = 51.2578125$

6.024 For the sequence $\dfrac{1}{7}, \dfrac{2}{7}, \dfrac{4}{7}, \dfrac{8}{7}, …$ write the recursive formula. Use the explicit formula to find a_8 and a_{15}.

The common ratio (r) is $a_2 \div a_1 = \dfrac{2}{7} \div \dfrac{1}{7} = \dfrac{2}{7} \times \dfrac{7}{1} = \dfrac{14}{7} = 2$ Recursive formula: $a_n = a_{n-1} \times 2$

To find the 8th term (a_8):

$a_n = a_1 \cdot r^{n-1}$

$a_8 = \dfrac{1}{7} \cdot (2)^{8-1}$

$a_8 = \dfrac{1}{7} \cdot (2)^7 = \dfrac{128}{7}$

To find the 15th term (a_{15}):

$a_n = a_1 \cdot r^{n-1}$

$a_{15} = \dfrac{1}{7} \cdot (2)^{15-1}$

$a_{15} = \dfrac{1}{7} \cdot (2)^{14} = \dfrac{16384}{7}$

6.025 Find the next three terms of the sequence 27, –9, 3, –1, … using the recursive formula.

The common ratio (r) is $\dfrac{a_2}{a_1} = \dfrac{-9}{27} = \dfrac{-1}{3}$ Recursive formula: $a_n = a_{n-1} \times \dfrac{-1}{3}$

The next three terms would be:

$a_5 = a_4 \times \dfrac{-1}{3} = -1 \times \dfrac{-1}{3} = \dfrac{1}{3}$ | $a_6 = a_5 \times \dfrac{-1}{3} = \dfrac{1}{3} \times \dfrac{-1}{3} = \dfrac{-1}{9}$ | $a_7 = a_6 \times \dfrac{-1}{3} = \dfrac{-1}{9} \times \dfrac{-1}{3} = \dfrac{1}{27}$

6.026 For the sequence $-\dfrac{1}{64}, \dfrac{1}{16}, -\dfrac{1}{4}, 1, …$ write the recursive formula. Use the explicit formula to find a_{12}.

The common ratio (r) is $a_2 \div a_1 = \dfrac{1}{16} \div \left(-\dfrac{1}{64}\right) = \dfrac{1}{16} \times \left(-\dfrac{64}{1}\right) = -\dfrac{64}{16} = -4$

Recursive formula: $a_n = a_{n-1} \times (-4)$

To find the 12th term (a_{12}):

$a_n = a_1 \cdot r^{n-1}$

$a_{12} = -\dfrac{1}{64} \cdot (-4)^{12-1} = -\dfrac{1}{64} \cdot (-4)^{11} = 65536$

PRACTICE PROBLEMS

In each of the following sequences, find the indicated terms and write the recursive formulas:

6.027 $\frac{1}{2}, -\frac{1}{3}, \frac{2}{9}, -\frac{4}{27}, ...$ Find a_6 and a_9.

6.028 $\frac{1}{4}, \frac{1}{5}, \frac{4}{25}, \frac{16}{125}, ...$ Find the next three terms.

6.029 $1, 1.1, 1.21, 1.331, ...$ Find a_6 and a_8.

6.030 $-100, 20, -4, 0.8, ...$ Find a_7.

GEOMETRIC SERIES

Let's revisit the geometric sequence 2, 6, 18, 54, … Adding up all those terms would give us the infinite **geometric series** 2+6+18+54+… The sum of this infinite series would be ∞ (undefined), but we can find the sum of a finite portion of it, maybe the first 5 terms, or 10 terms etc. with the explicit formula:

$$S_n = a_1 \cdot \left(\frac{1-r^n}{1-r} \right)$$ where

S_n is the sum of the first n terms
a_1 is the first term
r is the common ratio
n is the number of terms to be added up

Using this formula let's calculate S_{10} for the above geometric series.

We see that a_1 is 2, n is 10 (for S_{10}), and the common ratio is $r = \frac{a_2}{a_1} = \frac{6}{2} = 3$

Plug these things into the above formula for S_n:

$$S_{10} = 2 \cdot \left(\frac{1-3^{10}}{1-3} \right)$$

$$S_{10} = 2 \cdot \left(\frac{-59048}{-2} \right)$$

$$S_{10} = 59048$$

Please note that the sum of an infinite geometric series is *not* always undefined (+∞ or −∞). It is always undefined for infinite *arithmetic* series. For geometric series, as we'll see in the next section, it varies from case to case. In this section, however, let's solve problems on *finite* geometric series only.

6.031 Find the sum of the first nine terms (S_9) in the series $\frac{1}{2} + \frac{3}{2} + \frac{9}{2} + \frac{27}{2} + ...$

We see that a_1 is $\frac{1}{2}$, n is 9 (for S_9), and the common ratio is $r = a_2 \div a_1 = \frac{3}{2} \div \frac{1}{2} = \frac{3}{2} \times \frac{2}{1} = \frac{6}{2} = 3$

$$S_n = a_1 \cdot \left(\frac{1-r^n}{1-r} \right)$$

$$S_9 = \frac{1}{2} \cdot \left(\frac{1-3^9}{1-3} \right)$$

$$S_9 = \frac{1}{2} \cdot \left(\frac{-19682}{-2} \right) = \frac{9841}{2} = 4920.5$$

6.032 Find S_7 for the sequence 20, –2, 0.2, –0.02, …

We see that a_1 is 20, n is 7 (for S_7), and the common ratio is $r = \dfrac{a_2}{a_1} = \dfrac{-2}{20} = \dfrac{-1}{10} = -0.1$

$$S_n = a_1 \cdot \left(\frac{1 - r^n}{1 - r} \right)$$

$$S_7 = 20 \cdot \left(\frac{1 - (-0.1)^7}{1 - (-0.1)} \right)$$

$$S_7 = 20 \cdot \left(\frac{1.0000001}{1.1} \right) = 18.18182$$

6.033 Find S_{10} for the series $\dfrac{1}{16} + \dfrac{1}{8} + \dfrac{1}{4} + \dfrac{1}{2} + ...$

We see that a_1 is $\dfrac{1}{16}$, n is 10, and the common ratio is $r = a_2 \div a_1 = \dfrac{1}{8} \div \dfrac{1}{16} = \dfrac{1}{8} \times \dfrac{16}{1} = \dfrac{16}{8} = 2$

$$S_n = a_1 \cdot \left(\frac{1 - r^n}{1 - r} \right)$$

$$S_{10} = \frac{1}{16} \cdot \left(\frac{1 - 2^{10}}{1 - 2} \right)$$

$$S_{10} = \frac{1}{16} \cdot \left(\frac{-1023}{-1} \right) = \frac{1023}{16} = 63.9375$$

6.034 Find S_8 for the sequence $\dfrac{1}{4^3}, \dfrac{1}{4^2}, \dfrac{1}{4}, 1, ...$

We see that a_1 is $\dfrac{1}{4^3}$, n is 8, and the common ratio is $r = a_2 \div a_1 = \dfrac{1}{4^2} \div \dfrac{1}{4^3} = \dfrac{1}{4^2} \times \dfrac{4^3}{1} = \dfrac{4^3}{4^2} = 4$

$$S_n = a_1 \cdot \left(\frac{1 - r^n}{1 - r} \right)$$

$$S_8 = \frac{1}{4^3} \cdot \left(\frac{1 - 4^8}{1 - 4} \right)$$

$$S_8 = \frac{1}{64} \cdot \left(\frac{-65535}{-3} \right) = \frac{21845}{64} = 341.328125$$

PRACTICE PROBLEMS

6.035 Find S_{200} for the series $(-7) + 7 + (-7) + 7 + ...$

6.036 Calculate S_8 for the sequence $\dfrac{1}{125}, \dfrac{1}{25}, \dfrac{1}{5}, 1, ...$

6.037 Evaluate S_6 for the series $3 + 3^2 + 3^3 + 3^4 + ...$

6.038 Compute S_{13} for the sequence 3072, –1536, 768, –384, …

CONVERGENT AND DIVERGENT SERIES

If you start to add up the terms of an infinite sequence, and the result heads toward either positive or negative infinity, then its series is said to be **divergent** because it *diverges* away from zero in either positive or negative direction. That makes *all* arithmetic series divergent.

If the result of addition of the terms approaches a specific number (maybe zero, maybe some other number), then the series is **convergent** because it *converges* upon that number. Some geometric series are convergent; some divergent.

Let's examine the following infinite geometric series: $1 + \frac{1}{2} + \frac{1}{4} + \frac{1}{8} + ...$ Each term is half the previous term. Meaning, the common ratio is $\frac{1}{2}$. Is the sum of all these terms heading toward positive infinity? I don't know, but we can find out by adding all of its infinitely many terms! Can that really be done? Actually, yes. Not by adding them one by one, but by using the formula for S_n. Here's how:

We know that a_1 is 1, r is $\frac{1}{2}$, and n is ∞.

$$S_n = a_1 \cdot \left(\frac{1 - r^n}{1 - r} \right)$$

$$S_\infty = 1 \cdot \left(\frac{1 - \left(\frac{1}{2} \right)^\infty}{1 - \frac{1}{2}} \right)$$

Notice that $\left(\frac{1}{2} \right)^\infty = \frac{1}{2} \times \frac{1}{2} \times \frac{1}{2} \times ...$ infinitely many times. That's a fraction so tiny, it's practically zero.

$$S_\infty = 1 \cdot \left(\frac{1 - 0}{1 - \frac{1}{2}} \right) = 1 \cdot \left(\frac{1}{\frac{1}{2}} \right) = 1 \cdot 1 \cdot \frac{2}{1} = 2$$

Since the sum of the infinitely many terms converges upon 2, a specific number, the series is *convergent*.

Another way to tell whether a geometric series is convergent or divergent is by its common ratio (r). If the value of r is within the interval $(-1, 1)$ then the series is convergent.

6.039 Is the series in example 6.031 convergent or divergent? If convergent, what is its sum?

The series $\frac{1}{2} + \frac{3}{2} + \frac{9}{2} + \frac{27}{2} + ...$ from example 6.031 has a common ratio (r) equal to 3 which is not within the interval $(-1, 1)$. The series is therefore divergent.

6.040 Is the series in example 6.032 convergent? If yes, what value does it converge upon?

The series $20 + (-2) + 0.2 + (-0.02) + ...$ from example 6.032 has a common ratio (r) equal to -0.1 which is within the interval $(-1, 1)$. The series is therefore convergent. It converges upon its sum as calculated below:

$$S_n = a_1 \cdot \left(\frac{1 - r^n}{1 - r} \right) \quad \text{where } a_1 \text{ is 20, } r \text{ is } -0.1, \text{ and } n \text{ is } \infty.$$

$$S_\infty = 20 \cdot \left(\frac{1-(-0.1)^\infty}{1-(-0.1)} \right)$$

$$S_\infty = 20 \cdot \left(\frac{1-0}{1+0.1} \right) = 20 \cdot \left(\frac{1}{1.1} \right) = \frac{20}{1.1} = \frac{200}{11} = 18.\overline{18}$$

6.041 Is the geometric series $54+18+6+2+...$ convergent or divergent? If convergent, find its sum.

We see that a_1 is 54, and n is ∞. The common ratio is $r = \frac{a_2}{a_1} = \frac{18}{54} = \frac{1}{3}$ which is within the interval $(-1,1)$. The series is therefore convergent. Its sum is calculated below:

$$S_n = a_1 \cdot \left(\frac{1-r^n}{1-r} \right)$$

$$S_\infty = 54 \cdot \left(\frac{1-\left(\frac{1}{3}\right)^\infty}{1-\frac{1}{3}} \right)$$

$$S_\infty = 54 \cdot \left(\frac{1-0}{\frac{2}{3}} \right) = 54 \cdot \left(\frac{1}{\frac{2}{3}} \right) = 54 \cdot 1 \cdot \frac{3}{2} = 81$$

PRACTICE PROBLEMS

6.042 Would a series made with example 6.038 be convergent or divergent? If convergent, find its sum.

6.043 Is the series $\frac{1}{5} + \frac{1}{5^2} + \frac{1}{5^3} + \frac{1}{5^4} + ...$ convergent? If yes, what value does it converge upon?

CHAPTER EXERCISES

6.044 Given the arithmetic sequence $2.5,\ 2.9,\ 3.3,\ 3.7,\ ...$ find a_{25}. Also write the recursive formula.

6.045 In the arithmetic sequence $99,\ 84,\ 79,\ 64,\ ...$ find a_{20}. Also mention the recursive formula.

6.046 For the arithmetic series $7+10+13+16+...$ find the sum of the first 50 terms, written as S_{50}.

6.047 Find S_{15} for the arithmetic sequence $-1,\ -7,\ -13,\ -19,\ ...$

6.048 Find a_9 in the geometric sequence $6,\ 12,\ 24,\ 48,\ ...$ Also write the recursive formula.

6.049 In the geometric sequence $63,\ -21,\ 7,\ -\frac{7}{3},\ ...$ find a_7. Also write the recursive formula.

6.050 For the geometric series $30+(-3)+0.3+(-0.03)+...$ find S_8.

6.051 Find S_5 for the geometric series $1792+(-448)+112+(-28)+...$

6.052 Is the geometric series $\frac{1}{3} - \frac{1}{3^2} + \frac{1}{3^3} - \frac{1}{3^4} + ...$ convergent or divergent? If convergent, find its sum.

6.053 Is the geometric series $\frac{1}{6} + \frac{1}{6^2} + \frac{1}{6^3} + \frac{1}{6^4} + ...$ convergent? If yes, find the value it converges upon

CHAPTER 7
COMPLEX NUMBERS

In Chapter 1 (page 2) of this book, I made a passing reference to the term *imaginary numbers*. Now let's pursue an actual discussion of it. We know that $\sqrt{25}$ can be either 5 or –5, because $5\times5=25$ and $(-5)\times(-5)=25$ as well. But what would $\sqrt{-25}$ be? Similarly, what is $\sqrt{-36}$? What about $\sqrt{-100}$? As you can tell, square roots of negative numbers don't exist in the real world (and thus cannot be expressed on the *real number line*). Such numbers can only be imagined, and are therefore called **imaginary numbers**.

If you are curious about why we are studying this topic, let me assure you that imaginary numbers have extensive applications in the fields of electrical engineering, quantum physics and advanced trigonometry. We will, however, not go too deep into the topic; we will just scratch the surface.

THE IMAGINARY UNIT: i

Imaginary numbers are built on what's called the **imaginary unit**, designated by the letter i, and assigned the value $\sqrt{-1}$. If you need a refresher on how square roots (radicals) work in general, refer page 18. Now let's revisit the numbers $\sqrt{-25}$, $\sqrt{-36}$, and $\sqrt{-100}$ to see how we can express them in terms of the imaginary unit, i.

$$\sqrt{-25}$$
$$\sqrt{25\times(-1)}$$
$$5\sqrt{-1}$$
$$5i$$

$$\sqrt{-36}$$
$$\sqrt{36\times(-1)}$$
$$6\sqrt{-1}$$
$$6i$$

$$\sqrt{-100}$$
$$\sqrt{100\times(-1)}$$
$$10\sqrt{-1}$$
$$10i$$

Notice that in each case above, $\sqrt{-1}$ was rewritten as i. It's important to remember that when the imaginary unit is squared, we get –1, not 1. Meaning, $i^2=-1$. We will use this equivalence quite frequently in the later sections of this chapter.

COMPLEX NUMBER: $a + bi$

A complex number is of the form $a+bi$ where a and b are real numbers. Some examples of complex numbers are $2+3i$, $8-12i$, $-5+3i$ etc. Within a complex number, the real component is a, whereas the imaginary component is bi. A complex number resembles a polynomial written in i instead of x.

ADDITION/SUBTRACTION OF COMPLEX NUMBERS

Complex numbers can be added or subtracted exactly along the lines of polynomials. All you have to do is combine the *like terms*. If you need a refresher on that, visit page 25.

Consider the operation $2+3i$ plus $8-12i$. Combining the *like terms*, that would be $2+8$ which is 10, and $3i+(-12i)$ which is $-9i$. The answer is $10-9i$. Similarly, what is $5-i$ minus $4+3i$? That would be $5-4$ which is 1, and $-i-3i$ which is $-4i$. The answer is $1-4i$.

7.001 $(7+4i)+(8+i)$

Regroup the like terms and rewrite.

$$\overbrace{7+8} \quad \overbrace{+4i+i}$$

Please note that the $\overbrace{\quad}$ wave shown above the text is not a mathematical symbol. It's just my way of highlighting that those terms have been grouped.

$15+5i$

7.002 $(-2i+3)+(-3-6i)$

Combine the like terms.

$$\overbrace{3+(-3)} \quad \overbrace{-2i+(-6i)}$$
$3-3 \quad -2i-6i$
$0-8i$
$-8i$

Writing the 0 is optional. You can keep it if you want, or you can remove it.

7.003 $(22+i)-(3+13i)$

Combine the like terms.

$$\overbrace{22-3} \quad \overbrace{+i-13i}$$
$19-12i$

7.004 $(6+5i)+(4-5i)$

Combine the like terms.

$$\overbrace{6+4} \quad \overbrace{+5i-5i}$$
$10+0i$
10

7.005 $(16+15i)-7i$

Combine the like terms.

$$16 \quad \overbrace{+15i-17i}$$
$16-2i$

7.006 $(i-21)-(21-i)$

The answer is *not* zero, as you'll see.

$$\overbrace{i-(-i)} \quad \overbrace{-21-21}$$
$i+i \quad -42$
$2i-42$
$-42+2i$

PRACTICE PROBLEMS

7.007 $(6+2i)-(18-i)$
7.008 $(-1+4i)+(17-3i)$
7.009 $(2i-14)+(20-2i)$
7.010 $(7+6i)-(-5i+7)$
7.011 $(-i+11)+(-2i)$
7.012 $(-9+3i)-(10-7i)$

MULTIPLICATION OF COMPLEX NUMBERS

Recall from page 28 that when we multiplied two polynomials, we distributed each term of the first polynomial onto the terms of the second polynomial. We will do the exact same thing when multiplying two complex numbers as well, keeping in mind that $i^2 = -1$. Meaning, we will replace i^2 with -1 every time we come across it.

7.013 $(2+5i)(8-3i)$

Distribute 2 onto $(8-3i)$. Then distribute $5i$ onto $(8-3i)$.

$2 \cdot 8 + 2 \cdot (-3i) + (5i) \cdot 8 + (5i) \cdot (-3i)$

$16 - 6i + 40i - 15i^2$

Replace i^2 with -1.

$16 + 34i - 15(-1)$

$16 + 34i + 15$

$31 + 34i$

7.014 $(6i-3)(-9+i)$

Distribute $6i$ over $(-9+i)$. Then distribute -3 over $(-9+i)$.

$(6i) \cdot (-9) + (6i) \cdot (i) - 3 \cdot (-9) - 3 \cdot (i)$

$-54i + 6i^2 + 27 - 3i$

Replace i^2 with -1.

$-57i + 6(-1) + 27$

$-57i - 6 + 27$

$-57i + 21$

$21 - 57i$

7.015 $(4+2i)(1+i)$

Distribute 4 onto $(1+i)$. Then distribute $2i$ onto $(1+i)$.

$4 \cdot 1 + 4 \cdot i + (2i) \cdot 1 + (2i) \cdot i$

$4 + 4i + 2i + 2i^2$

Replace i^2 with -1.

$4 + 6i + 2(-1)$

$4 + 6i - 2$

$2 + 6i$

7.016 $(5+4i)(-4i+5)$

Distribute 5 over $(-4i+5)$. Then distribute $4i$ over $(-4i+5)$.

$5 \cdot (-4i) + 5 \cdot 5 + (4i) \cdot (-4i) + (4i) \cdot 5$

$-20i + 25 - 16i^2 + 20i$

Replace i^2 with -1. The $20i$ and $-20i$ cancel each other.

$25 - 16(-1)$

$25 + 16$

41

Multiplication of two complex numbers *can* result in a real number.

PRACTICE PROBLEMS

7.017 $(1-8i)(5+3i)$

7.018 $(-7i+6)(2-5i)$

7.019 $(4+2i)(-4+2i)$

7.020 $(3+25i)(1-i)$

DIVISION OF COMPLEX NUMBERS

Imagine dividing $5-2i$ by $3+4i$. We could write it as $\frac{5-2i}{3+4i}$. Division would be possible if the denominator $3+4i$ could somehow be converted into a real number. That can be achieved by multiplying it by its *complex conjugate*. A **complex conjugate** is basically the original complex number with its imaginary component rewritten with the sign flipped. For example, the complex conjugate of $5+3i$ would be $5-3i$, that of $-12+i$ would be $-12-i$, while that of $8i$ would be $-8i$.

So to achieve the division $\frac{5-2i}{3+4i}$, multiply the numerator and denominator by the complex conjugate of the denominator, which turns out to be $3-4i$.

$$\frac{5-2i}{3+4i}\times\frac{3-4i}{3-4i}$$

Since this resembles multiplication of rational expressions (refer page 40), multiply the numerators and denominators straight across.

$$\frac{(5-2i)\cdot(3-4i)}{(3+4i)\cdot(3-4i)}$$

$$\frac{15-20i-6i+8i^2}{9-12i+12i-16i^2}$$

Replace i^2 with -1.

$$\frac{15-26i+8(-1)}{9-16(-1)}$$

$$\frac{15-26i-8}{9+16}$$

$$\frac{7-26i}{25}$$

$$\frac{7}{25}-\frac{16}{25}i$$

This is the final answer, expressed in the form $a+bi$.

7.021 $\quad \dfrac{3+i}{2-i}$

The complex conjugate of the denominator is $2+i$. Multiply the numerator and denominator by it.

$$\frac{3+i}{2-i}\times\frac{2+i}{2+i}$$

Multiply the numerators and denominators straight across.

$$\frac{(3+i)\cdot(2+i)}{(2-i)\cdot(2+i)}$$

$$\frac{6+3i+2i+i^2}{4+2i-2i-i^2}$$

Replace i^2 with -1.

$$\frac{6+5i-1}{4-(-1)}$$

$$\frac{5+5i}{5}$$

$$\frac{5}{5}+\frac{5}{5}i$$

$$1+i$$

7.022 $\quad \dfrac{-5+6i}{1+3i}$

Multiply the numerator and denominator by the complex conjugate of the denominator, $1-3i$.

$$\frac{-5+6i}{1+3i}\times\frac{1-3i}{1-3i}$$

Multiply straight across.

$$\frac{(-5+6i)\cdot(1-3i)}{(1+3i)\cdot(1-3i)}$$

$$\frac{-5+15i+6i-18i^2}{1-3i+3i-9i^2}$$

Replace i^2 with -1.

$$\frac{-5+21i-18(-1)}{1-9(-1)}$$

$$\frac{-5+21i+18}{1+9}$$

$$\frac{13+21i}{10}$$

$$\frac{13}{10}+\frac{21}{10}i$$

7.023 $\dfrac{9i-7}{3i+2}$

The complex conjugate of the denominator is *not* $3i-2$, but $-3i+2$. Multiply the numerator and denominator by it.

$$\dfrac{9i-7}{3i+2} \times \dfrac{-3i+2}{-3i+2}$$

Multiply straight across.

$$\dfrac{(9i-7)\cdot(-3i+2)}{(3i+2)\cdot(-3i+2)}$$

$$\dfrac{-27i^2+18i+21i-14}{-9i^2+6i-6i+4}$$

Replace i^2 with -1.

$$\dfrac{-27(-1)+39i-14}{-9(-1)+4}$$

$$\dfrac{27+39i-14}{9+4}$$

$$\dfrac{13+39i}{13}$$

$$\dfrac{13}{13}+\dfrac{39}{13}i$$

$$1+3i$$

7.024 $\dfrac{6-8i}{2i}$

Multiply the numerator and denominator by the complex conjugate of the denominator, $-2i$.

$$\dfrac{6-8i}{2i} \times \dfrac{-2i}{-2i}$$

Multiply straight across.

$$\dfrac{(6-8i)\cdot(-2i)}{(2i)\cdot(-2i)}$$

$$\dfrac{-12i+16i^2}{-4i^2}$$

Replace i^2 with -1.

$$\dfrac{-12i+16(-1)}{-4(-1)}$$

$$\dfrac{-12i-16}{4}$$

$$\dfrac{-12}{4}i-\dfrac{16}{4}$$

$$-3i-4$$

$$-4-3i$$

PRACTICE PROBLEMS

7.025 $\dfrac{1+4i}{5-7i}$

7.026 $\dfrac{9-36i}{-9i}$

7.027 $\dfrac{2i-6}{-3+i}$

7.028 $\dfrac{-7+4i}{10-i}$

POWERS OF i

We know that $i=\sqrt{-1}$, which means $i^2=-1$. Let's see what happens when i is raised to higher powers (higher exponents), as in i^3, i^4, i^5 etc. (Review *laws of exponents* from page 43 if need be.)

i

$i^2=-1$

$i^3=i^2\cdot i=(-1)\cdot i=-i$

$i^4=i^2\cdot i^2=(-1)\cdot(-1)=1$

$i^5=i^4\cdot i=1\cdot i=i$

$i^6=i^4\cdot i^2=1\cdot(-1)=-1$

$i^7=i^6\cdot i=(-1)\cdot i=-i$

$i^8=i^4\cdot i^4=1\cdot 1=1$

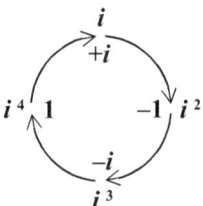

We see a pattern here: i, -1, $-i$, 1, i, -1, $-i$, 1,... Meaning, every fourth power of i results in the same value. So, to evaluate i raised to any number, break that number into a multiple of 4 plus the remainder. See where it fits into the pattern. For example, $i^{18}=i^{16+2}=i^{16}\cdot i^2=1\cdot(-1)=-1$

The picture above on the right shows the cyclic nature of the powers of i.

7.029 $(i^3)^4$

$i^{3\times4}$ Refer page 43, law 3.

i^{12}

1   because $i^{\text{any multiple of }4}=i^4=1$

7.030 $i^{11}(i^3+3i^4)$

Distribute and simplify.

$i^{11+3}+3i^{11+4}$ Refer page 43, law 1.

$i^{14}+3i^{15}$

$i^{12+2}+3i^{12+3}$

$i^{12}\cdot i^2+3i^{12}\cdot i^3$ Refer page 43, law 1.

$1\cdot(-1)+3\cdot1\cdot i^3$

$-1+3i^3$

$-1+3i^2\cdot i$

$-1+3(-1)\cdot i$

$-1-3i$

7.031 i^{41}

i^{40+1}

$i^{40}\cdot i^1$ Refer page 43, law 1.

$1\cdot i$ because $i^{40}=i^4=1$

i

7.032 $(2-i^7)(i^4-3)$

$2i^4-6-i^{7+4}+3i^7$ Refer page 43, law 1.

$2\cdot1-6-i^{11}+3i^7$

$2-6-i^{8+3}+3i^{4+3}$

$-4-i^8\cdot i^3+3i^4\cdot i^3$ Refer page 43, law 1.

$-4-1\cdot i^3+3\cdot1\cdot i^3$ because $i^8=i^4=1$

$-4-i^3+3i^3$

$-4+2i^3$

$-4+2i^2\cdot i$

$-4+2(-1)\cdot i$

$-4-2i$

PRACTICE PROBLEMS

7.033 $(2i^5)^3$ **7.034** i^{22} **7.035** $(i^3+i)(i^{10}-i^6)$ **7.036** $25i^6\cdot13i^7$

CHAPTER EXERCISES

7.037 $(16+11i)+(-4-4i)$

7.038 $(5-4i)-(-4i+7)$

7.039 $(6i+8)+(6-4i)$

7.040 $(5i+4)-(9i+4)$

7.041 $(2+i)(-8+3i)$

7.042 $(4+i)(i-4)$

7.043 $\dfrac{3-5i}{7+3i}$

7.044 $\dfrac{18i+12}{3i}$

7.045 $i^{13}-(i^5)^7$

7.046 $-6i(5i-i^3)$

APPENDIX A
SOLUTIONS TO PRACTICE PROBLEMS AND CHAPTER EXERCISES

CHAPTER 1: ARITHMETIC (PRE-ALGEBRA)

1.007
$21-(+3)$
$21-3$
18

1.008
$18+(-7)-(+2)$
$18-7-2$
$11-2$
9

1.009
$-1-(-6)+(-14)$
$-1+6-14$
$5-14$
-9

1.010
$4-(+4)-(-10)$
$4-4+10$
$0+10$
10

1.011
$-11-(-17)-9$
$-11+17-9$
$6-9$
-3

1.012
$31+(-26)-(-3)$
$31-26+3$
$5+3$
8

1.013
$-16-(-8)+21$
$-16+8+21$
$-8+21$
13

1.014
$-6+(-16)-(-26)$
$-6-16+26$
$-22+26$
4

1.019
$(-3)\times(-9)\times(-1)$
$-3\times9\times1$
-27×1
-27

1.020
$6\times(-5)\times(-2)$
$6\times5\times2$
30×2
60

1.021
$(-1)\times(-5)\times2\times(-1)$
$-1\times5\times2\times1$
$-5\times2\times1$
-10×1
-10

1.022
$(-8)\times(-3)\times2$
$8\times3\times2$
24×2
48

1.023
$4\times(-3)\times2$
$-4\times3\times2$
-12×2
-24

1.024
$(-7)\times6$
-42

1.029
$\dfrac{-60}{12}$

The answer will be *negative*.
-5

1.030
$(-35)\div7$

The answer will be *negative*.
-5

1.031
$(-42)\div(-6)$

The answer will be *positive*.
7

1.032
$\dfrac{14}{(-2)}$

The answer will be *negative*.
-7

1.033
$\dfrac{(-80)}{(-4)}$

The answer will be *positive*.
20

1.034
$28\div(-4)$

The answer will be *negative*.
-7

1.040
$(-4)^3$

The answer will be *negative*.
$-4\times4\times4$
-64

1.041 -2^6

The answer will remain *negative* because there are no parentheses around 2.

$-2\times2\times2\times2\times2\times2$

-64

1.042 $(-2)^6$

The answer will be *positive*.

$2\times2\times2\times2\times2\times2$

64

1.043 $(-5)^3$

The answer will be *negative*.

$-5\times5\times5$

-125

1.044 -1^{50}

The answer will remain *negative* because there are no parentheses around 1.

-1

1.045 $(-1)^{807}$

The answer will be *negative*.

-1

1.051 $(\overbrace{-7+5})^3\times(-7)\div2$

$\overbrace{(-2)^3}\times(-7)\div2$

$\overbrace{-8\times(-7)}\div2$

$56\div2$

28

1.052 $\overbrace{-48\div12}\times7+3$

$\overbrace{-4\times7}+3$

$-28+3$

-25

1.053 $(\overbrace{100-49})\div3+\dfrac{-8}{2}$

$\overbrace{51\div3}+\overbrace{\dfrac{-8}{2}}$

$17+(-4)$

$17-4$

13

1.054 $\dfrac{(\overbrace{6+(-3)})^3}{9}$

$\dfrac{(\overbrace{6-3})^3}{9}$

$\dfrac{3^3}{9}$

$\dfrac{27}{9}$

3

1.055 $\overbrace{4-(-7)}\times(-6)+12$

Simplify only the *signs* first. Do not calculate anything in this step.

$\overbrace{4+7\times(-6)}+12$ Now calculate.

$\overbrace{4+(-42)}+12$

$4-42+12$

$-38+12$

-26

1.056 $6\times(\overbrace{4-5})^{207}+6$

$6\times\overbrace{(-1)^{207}}+6$

$\overbrace{6\times(-1)}+6$

$-6+6$

0

1.059 20 can be obtained in the following ways:

1×20

2×10

4×5

The factors of 20 are: 1, 2, 4, 5, 10, 20

12 can be obtained in the following ways:

1×12

2×6

3×4

The factors of 12 are: 1, 2, 3, 4, 6, 12

The GCF of 20 and 12 is 4.

1.060 45 can be obtained in the following ways:

1×45

3×15

5×9

The factors of 45 are: 1, 3, 5, 9, 15, 45

72 can be obtained in the following ways:

1×72

2×36

3×24

4×18

6×12

8×9

The factors of 72 are:

1, 2, 3, 4, 6, 8, 9, 12, 18, 24, 36, 72

The GCF of 45 and 72 is 9.

1.061 16 can be obtained in the following ways:

1×16

2×8

4×4

The factors of 16 are: 1, 2, 4, 8, 16

40 can be obtained in the following ways:
  1×40
  2×20
  4×10
  5×8
The factors of 40 are:1,2,4,5,8,10,20,40

12 can be obtained in the following ways:
  1×12
  2×6
  3×4
The factors of 12 are: 1, 2, 3, 4, 6, 12
The GCF of 16, 40 and 12 is 4.

1.062 35 can be obtained in the following ways:
  1×35
  5×7
The factors of 35 are: 1, 5, 7, 35

45 can be obtained in the following ways:
  1×45
  3×15
  5×9
The factors of 45 are: 1, 3, 5, 9, 15, 45

55 can be obtained in the following ways:
  1×55
  5×11
The factors of 55 are: 1, 5, 11, 55
The GCF of 35, 45 and 55 is 5.

1.065 Multiples of 15 are: 15, 30, 45, 60, 75, …
Multiples of 25 are: 25, 50, 75, …
Lowest number on both lists is 75, and is therefore the LCM.

1.066 Multiples of 12 are: 12, 24, 36, 48, …
Multiples of 16 are: 16, 32, 48, …
Lowest number on both lists is 48, and is therefore the LCM.

1.067 Multiples of 20 are:
20, 40, 60, 80, 100, 120, …
Multiples of 30 are: 30, 60, 90, 120, …
Multiples of 40 are: 40, 80, 120, …
Lowest number on these lists is 120, and is therefore the LCM.

1.068 Multiples of 3 are:
3, 6, 9, 12, 15, 18, 21, 24, 27, …
Multiples of 9 are: 9, 18, 27, …
Multiples of 27 are: 27, 54, …
Lowest number on these lists is 27, and is therefore the LCM.

1.072 $\dfrac{25}{7}$

$7 \times 3 = 21$, the remainder being 4.

Thus, the answer is $3\dfrac{4}{7}$.

1.073 $\dfrac{47}{20}$

$20 \times 2 = 40$, the remainder being 7.

Thus, the answer is $2\dfrac{7}{20}$.

1.074 $\dfrac{9}{2}$

$2 \times 4 = 8$, the remainder being 1.

Thus, the answer is $4\dfrac{1}{2}$.

1.075 $\dfrac{16}{3}$

$3 \times 5 = 15$, the remainder being 1.

Thus, the answer is $5\dfrac{1}{3}$.

1.078 3×7 gives 21
$21 + 2$ gives 23

Answer $= \dfrac{23}{3}$

$7\dfrac{2}{3}$ *plus* / *times*

1.079 4×5 gives 20
$20 + 1$ gives 21

Answer $= \dfrac{21}{4}$

$5\dfrac{1}{4}$ *plus* / *times*

1.080 2×15 gives 30
$30 + 1$ gives 31

Answer $= \dfrac{31}{2}$

$15\dfrac{1}{2}$ *plus* / *times*

1.081 8×6 gives 48

48 + 3 gives 51

Answer = $\dfrac{51}{8}$

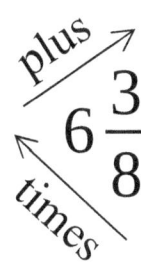

$6\dfrac{3}{8}$ *plus* *times*

1.088 $\dfrac{9}{7} + \dfrac{13}{7}$

Denominators are the same. Add the numerators.

$\dfrac{9+13}{7}$

$\dfrac{22}{7}$

$3\dfrac{1}{7}$

1.089 $\dfrac{5}{8} - \dfrac{3}{8}$

Denominators are the same. Subtract the numerators.

$\dfrac{5-3}{8}$

$\dfrac{2}{8}$

Reduce to *lowest terms* by dividing top and bottom by 2.

$\dfrac{1}{4}$

1.090 $\dfrac{16}{21} + \dfrac{2}{21} - \dfrac{11}{21}$

Denominators are the same. Combine the numerators.

$\dfrac{16+2-11}{21}$

$\dfrac{7}{21}$

Reduce to *lowest terms* by dividing top and bottom by 7.

$\dfrac{1}{3}$

1.091 $\dfrac{3}{7} + \dfrac{1}{4}$

Equalize the denominators, their LCM being 28.

$\dfrac{3\times4}{7\times4} + \dfrac{1\times7}{4\times7}$

$\dfrac{12}{28} + \dfrac{7}{28}$

$\dfrac{12+7}{28}$

$\dfrac{19}{28}$

The fraction is already in its *lowest terms*.

1.092 $\dfrac{1}{2} - \dfrac{1}{3} - \dfrac{1}{4}$

Equalize the denominators, their LCM being 12.

$\dfrac{1\times6}{2\times6} - \dfrac{1\times4}{3\times4} - \dfrac{1\times3}{4\times3}$

$\dfrac{6}{12} - \dfrac{4}{12} - \dfrac{3}{12}$

$\dfrac{-1}{12}$

1.093 $2\dfrac{1}{8} - 1\dfrac{3}{4}$

First, convert both fractions into improper fractions.

$\dfrac{8\times2+1}{8} - \dfrac{4\times1+3}{4}$

$\dfrac{17}{8} - \dfrac{7}{4}$

Equalize the denominators, their LCM being 8.

$\dfrac{17}{8} - \dfrac{7\times2}{4\times2}$

Notice that there was no need to raise the first fraction to *higher terms*.

$\dfrac{17}{8} - \dfrac{14}{8}$

$\dfrac{3}{8}$

1.097 $\dfrac{6}{7} \times \dfrac{2}{9}$

Reduce the 6 and the 9 by dividing by 3.

$\dfrac{\cancel{6}^{2}}{7} \times \dfrac{2}{\cancel{9}^{3}}$

$\dfrac{2\times2}{7\times3}$

$\dfrac{4}{21}$

1.098 $\dfrac{3}{8} \times \dfrac{5}{1} \times \dfrac{1}{8}$

No reductions or cross reductions are possible. So multiply straight across.

$$\dfrac{3 \times 5 \times 1}{8 \times 1 \times 8}$$

$$\dfrac{15}{64}$$

1.099 $4\dfrac{1}{2} \times \dfrac{21}{25} \times \dfrac{2}{7}$

First, convert the mixed fraction into an improper fraction.

$$\dfrac{2 \times 4 + 1}{2} \times \dfrac{21}{25} \times \dfrac{2}{7}$$

$$\dfrac{9}{2} \times \dfrac{21}{25} \times \dfrac{2}{7}$$

Reduce 2 and 2 by dividing by 2. Reduce 7 and 21 by dividing by 7.

$$\dfrac{9}{\cancel{2}^{1}} \times \dfrac{\cancel{21}^{3}}{25} \times \dfrac{\cancel{2}^{1}}{\cancel{7}^{1}}$$

$$\dfrac{9 \times 3 \times 1}{1 \times 25 \times 1}$$

$\dfrac{27}{25}$, optionally being $1\dfrac{2}{25}$.

1.103 $\dfrac{\dfrac{9}{8}}{\dfrac{3}{32}}$

Multiply the first fraction by the reciprocal of the second fraction.

$$\dfrac{9}{8} \times \dfrac{32}{3}$$

Cross reduce the fractions.

$$\dfrac{\cancel{9}^{3}}{\cancel{8}^{1}} \times \dfrac{\cancel{32}^{4}}{\cancel{3}^{1}}$$

$$\dfrac{3 \times 4}{1 \times 1}$$

$$\dfrac{12}{1}$$

$$12$$

1.104 $\dfrac{16}{81} \div \dfrac{8}{9}$

Multiply the first fraction by the reciprocal of the second fraction.

$$\dfrac{16}{81} \times \dfrac{9}{8}$$

Cross reduce the fractions.

$$\dfrac{\cancel{16}^{2}}{\cancel{81}^{9}} \times \dfrac{\cancel{9}^{1}}{\cancel{8}^{1}}$$

$$\dfrac{2 \times 1}{9 \times 1}$$

$$\dfrac{2}{9}$$

1.105 $\dfrac{1\dfrac{23}{40}}{\dfrac{7}{25}}$

Convert the mixed fraction into an improper fraction.

$$\dfrac{\dfrac{40 \times 1 + 23}{40}}{\dfrac{7}{25}}$$

$$\dfrac{\dfrac{63}{40}}{\dfrac{7}{25}}$$

Multiply the first fraction by the reciprocal of the second fraction.

$$\dfrac{63}{40} \times \dfrac{25}{7}$$

Cross reduce the fractions.

$$\dfrac{\cancel{63}^{9}}{\cancel{40}^{8}} \times \dfrac{\cancel{25}^{5}}{\cancel{7}^{1}}$$

$$\dfrac{9 \times 5}{8 \times 1}$$

$$\dfrac{45}{8}$$

$$5\dfrac{5}{8}$$

1.109 2.25

There are *two* decimal digits. Place them over 100 (because 100 has *two* zeroes). Keep the whole number component the same.

$$2\dfrac{25}{100}$$

Reduce the fractional component.

$$2\dfrac{1}{4}$$

1.110 −0.48

There are *two* decimal digits. Place them over 100 (because 100 has *two* zeroes).

$$-\frac{48}{100}$$

Reduce the fraction.

$$-\frac{24}{50}$$

$$-\frac{12}{25}$$

1.111 61.006

There are *three* decimal digits. Place them over 1000 (because 1000 has *three* zeroes). Keep the whole number component the same.

$$61\frac{006}{1000}$$

It's not a good practice to write 006 in the numerator. Instead, write 6.

$$61\frac{6}{1000}$$

Reduce the fractional component.

$$61\frac{3}{500}$$

1.112 −9.5

There is *one* decimal digit. Place it over 10 (because 10 has *one* zero).

$$-9\frac{5}{10}$$

Reduce the fractional component.

$$-9\frac{1}{2}$$

1.117 −28.8%

$$-\frac{28.8}{100}=-0.288$$

1.118 4%

$$\frac{4}{100}=\frac{4.0}{100}=\frac{00004.0}{100}=000.040=0.04$$

1.119 1.09%

$$\frac{1.09}{100}=\frac{00001.09}{100}=000.0109=0.0109$$

1.120 300%

$$\frac{300}{100}=\frac{300.0}{100}=3.000=3$$

1.121 −25%

$$-\frac{25}{100}=-\frac{1}{4}$$

1.122 53%

$$\frac{53}{100}$$

1.123 −10%

$$-\frac{10}{100}=-\frac{1}{10}$$

1.124 145%

$$\frac{145}{100}=\frac{29}{20}=1\frac{9}{20}$$

1.129 $-0.971\times100=-097.1=-97.1\%$

1.130 $3\times100=300\%$

1.131 $0.0153\times100=001.53=1.53\%$

1.132 $6.28\times100=628.0=628\%$

1.137 $1\frac{8}{9}=1+\frac{8}{9}=1+(8\div9)$

$$=1+0.888....=1+0.\bar{8}=1.\bar{8}$$

1.138 $\frac{4}{7}=4\div7=0.571428571428......$

$$=0.\overline{571428}$$

1.139 $-15\frac{5}{6}=-\left(15+\frac{5}{6}\right)=-(15+(5\div6))$

$$=-(15+0.83333.....)$$

$$=-(15+0.8\bar{3})=-15.8\bar{3}$$

1.140 $-3\frac{1}{4}=-\left(3+\frac{1}{4}\right)=-(3+(1\div4))$

$$=-(3+0.25)=-3.25$$

1.144 $-\frac{4}{5}\times100=-0.8\times100=-80\%$

1.145 $4\frac{2}{9}\times100=\left(4+\frac{2}{9}\right)\times100$

$$=(4+0.2222...)\times100=4.2222...\times100$$

$$=422.22....=422.\bar{2}\%$$

1.146 $\frac{11}{6}\times100=1.83333...\times100$

$$=183.333...=183.\bar{3}\%$$

1.150 $\sqrt{75}$

$$\sqrt{25\times3}$$

$\sqrt{25}$ is 5, while $\sqrt{3}$ is not an integer.

$$5\sqrt{3}$$

1.151 $\dfrac{\sqrt{35}}{\sqrt{7\times5}}$

Neither 7 nor 5 is a perfect square. So $\sqrt{35}$ can't be written with a smaller radicand.

1.152 $\dfrac{\sqrt{32}}{\sqrt{8\times4}}$
$\sqrt{4\times2\times4}$

Rearrange the numbers.
$\sqrt{4\times4\times2}$
$\sqrt{16\times2}$
$\sqrt{16}$ is 4, while $\sqrt{2}$ is not an integer.
$4\sqrt{2}$

1.157 $5\sqrt{7}-\sqrt{28}$

The radicands aren't the same. First, simplify $\sqrt{28}$.
$5\sqrt{7}-\sqrt{7\times4}$
$5\sqrt{7}-2\sqrt{7}$
$3\sqrt{7}$

1.158 $\sqrt{54}-\sqrt{24}+3\sqrt{6}$

The radicands aren't the same. First, simplify them.
$\sqrt{2\times27}-\sqrt{12\times2}+3\sqrt{3\times2}$
$\sqrt{2\times3\times9}-\sqrt{4\times3\times2}+3\sqrt{3\times2}$
$3\sqrt{2\times3}-2\sqrt{3\times2}+3\sqrt{3\times2}$

The radicands can't be simplified any further.
$3\sqrt{6}-2\sqrt{6}+3\sqrt{6}$
$4\sqrt{6}$

1.159 $3\sqrt{50}-4\sqrt{18}$

First, simplify the radicands.
$3\sqrt{25\times2}-4\sqrt{9\times2}$
$3\times5\sqrt{2}-4\times3\sqrt{2}$
$15\sqrt{2}-12\sqrt{2}$
$3\sqrt{2}$

1.160 $-7\sqrt{27}-5\sqrt{12}$

First, simplify the radicands.
$-7\sqrt{9\times3}-5\sqrt{4\times3}$
$-7\times3\sqrt{3}-5\times2\sqrt{3}$
$-21\sqrt{3}-10\sqrt{3}$
$-31\sqrt{3}$

1.164 $2\sqrt{18}\times18\sqrt{2}$

Multiply the coefficients, and combine the radicands under one radical sign.
$2\times18\sqrt{18\times2}$

$36\sqrt{36}$
36×6
216

1.165 $-4\sqrt{10}\times(-3)\sqrt{2}$

Multiply the coefficients, and combine the radicands under one radical sign.
$-4\times(-3)\sqrt{10\times2}$
$12\sqrt{5\times2\times2}$
$12\sqrt{5\times4}$
$12\times2\sqrt{5}$
$24\sqrt{5}$

1.166 $\sqrt{200}\times3\sqrt{8}$

The coefficient of $\sqrt{200}$ is not *nothing*, but 1.
$1\sqrt{200}\times3\sqrt{8}$

Multiply the coefficients, and combine the radicands under one radical sign.
$1\times3\sqrt{200\times8}$
$3\sqrt{2\times100\times8}$
$3\sqrt{2\times8\times100}$
$3\sqrt{16\times100}$
$3\times4\times10$
120

1.170 $2\sqrt{8}\div\sqrt{32}$

The coefficient of $\sqrt{32}$ is 1.
$\dfrac{2\sqrt{8}}{1\sqrt{32}}$

Write as a product of two separate fractions: one containing the coefficients and the other containing the radicals.
$\dfrac{2}{1}\times\dfrac{\sqrt{8}}{\sqrt{32}}$

Now combine the radicands.
$\dfrac{2}{1}\times\sqrt{\dfrac{8}{32}}$
$\dfrac{2}{1}\times\sqrt{\dfrac{1}{4}}$

Now break the radical again into two separate radicals. Then simplify.
$\dfrac{2}{1}\times\dfrac{\sqrt{1}}{\sqrt{4}}$
$\dfrac{2}{1}\times\dfrac{1}{2}$

$$\frac{2 \times 1}{1 \times 2}$$

$$\frac{2}{2}$$

$$1$$

1.171 $\frac{\sqrt{44}}{\sqrt{99}}$

Combine the radicals into one.

$$\sqrt{\frac{44}{99}}$$

Reduce 44 and 99 by 11.

$$\sqrt{\frac{4}{9}}$$

Break into two radicals. Then simplify.

$$\frac{\sqrt{4}}{\sqrt{9}}$$

$$\frac{2}{3}$$

1.172 $3\sqrt{75} \div 5\sqrt{27}$

$$\frac{3\sqrt{75}}{5\sqrt{27}}$$

$$\frac{3\sqrt{25 \times 3}}{5\sqrt{9 \times 3}}$$

$$\frac{3 \times 5\sqrt{3}}{5 \times 3\sqrt{3}}$$

$$\frac{15\sqrt{3}}{15\sqrt{3}}$$

$$1$$

1.173 $-21+(-13)-(+6)-(-2)$

$$-21-13-6+2$$

$$-34-6+2$$

$$-40+2$$

$$-38$$

1.174 $-4 \times (-5) \times 2$

$$20 \times 2$$

$$40$$

1.175 $\frac{-16}{-4}$

The answer will be *positive*.

$$4$$

1.176 $\overbrace{39 \div (-13)} \times 8$

Contains multiplication and division only.
Perform the operations from left to right.

$$-3 \times 8$$

$$-24$$

1.177 $\left(\frac{\overbrace{6-(-4)}}{5}\right)^2 + \overbrace{6 \times (-2)}$

$$\left(\frac{6+4}{5}\right)^2 + (-12)$$

$$\left(\frac{10}{5}\right)^2 - 12$$

$$2^2 - 12$$

$$4 - 12$$

$$-8$$

1.178 $\overbrace{-11+(-8)} \div 2 \times (-3)$

First simplify the sign.

$$-11 - \overbrace{8 \div 2} \times (-3)$$

Now, over the stretch involving multiplication and division, perform the calculations from left to right.

$$-11 - \overbrace{4 \times (-3)}$$

$$-11 - (-12)$$

$$-11 + 12$$

$$1$$

1.179 36 can be obtained in the following ways:

$$1 \times 36$$
$$2 \times 18$$
$$3 \times 12$$
$$4 \times 9$$
$$6 \times 6$$

The factors of 36 are:
1, 2, 3, 4, 6, 9, 12, 18, 36

20 can be obtained in the following ways:
$$1 \times 20$$
$$2 \times 10$$
$$4 \times 5$$

The factors of 20 are: 1, 2, 4, 5, 10, 20

The GCF of 36 and 20 is 4.

1.180 Multiples of 12 are:
12, 24, 36, 48, 60, …

Multiples of 15 are:
15, 30, 45, 60, …

The number 60 is the *lowest* number common to both lists, and is the LCM of 12 and 15.

1.181 3 times 5 gives 15, the remainder being 2. Thus, the answer is $5\frac{2}{3}$.

1.182

5 × 9 gives 45
45 + 2 gives 47

Answer = $\frac{47}{5}$

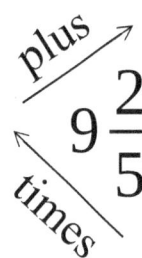

1.183 $\frac{1}{3} - \frac{5}{4} + \frac{7}{8}$

Equalize the denominators, their LCM being 24.

$$\frac{1\times8}{3\times8} - \frac{5\times6}{4\times6} + \frac{7\times3}{8\times3}$$

$$\frac{8}{24} - \frac{30}{24} + \frac{21}{24}$$

$$\frac{8-30+21}{24}$$

$$\frac{-22+21}{24}$$

$$\frac{-1}{24}$$

1.184 $2\frac{5}{6} + 10\frac{1}{3} - 6\frac{1}{2}$

First, convert mixed fractions into improper fractions.

$$\frac{6\times2+5}{6} + \frac{3\times10+1}{3} - \frac{2\times6+1}{2}$$

$$\frac{17}{6} + \frac{31}{3} - \frac{13}{2}$$

Equalize their denominators, their LCM being 6.

$$\frac{17}{6} + \frac{31\times2}{3\times2} - \frac{13\times3}{2\times3}$$

$$\frac{17}{6} + \frac{62}{6} - \frac{39}{6}$$

$$\frac{17+62-39}{6}$$

$$\frac{40}{6}$$

$$\frac{20}{3}$$

$6\frac{2}{3}$

1.185 $\frac{27}{34} \times 5\frac{2}{3}$

First, convert the mixed fraction into an improper fraction.

$$\frac{27}{34} \times \frac{3\times5+2}{3}$$

$$\frac{27}{34} \times \frac{17}{3}$$

Reduce 34 with 17, and 27 with 3.

$$\frac{\cancel{27}^{9}}{\cancel{34}^{2}} \times \frac{\cancel{17}^{1}}{\cancel{3}^{1}}$$

$$\frac{9}{2} \times \frac{1}{1}$$

$$\frac{9\times1}{2\times1}$$

$$\frac{9}{2}$$

$$4\frac{1}{2}$$

1.186 $7\frac{1}{6} \div \frac{22}{3}$

First, convert the mixed fraction into an improper fraction.

$$\frac{6\times7+1}{6} \div \frac{22}{3}$$

$$\frac{43}{6} \div \frac{22}{3}$$

First fraction times the reciprocal of the second fraction.

$$\frac{43}{\cancel{6}^{2}} \times \frac{\cancel{3}^{1}}{22}$$

$$\frac{43\times1}{2\times22}$$

$$\frac{43}{44}$$

1.187 −0.84 has *two* decimal digits. Place them over 100 (because 100 has *two* zeroes).

$$-\frac{84}{100}$$

Reduce both numbers by 4.

$$-\frac{21}{25}$$

1.188 $28\% = \dfrac{28}{100} = \dfrac{28.0}{100} = 0.28$

1.189 $\dfrac{5}{4} = 1.25 = 1.25 \times 100\% = 125\%$

1.190 $65\% = \dfrac{65}{100} = \dfrac{13}{20}$

1.191 $\dfrac{11}{12} = \dfrac{11}{12} \times 100 = 0.91666.... \times 100$

$= 91.666.... = 91.\overline{6}\%$

1.192 $\sqrt{84}$

$\sqrt{2 \times 42}$

$\sqrt{2 \times 2 \times 21}$

$\sqrt{2 \times 2 \times 7 \times 3}$

$\sqrt{4 \times 7 \times 3}$

$\sqrt{4}$ is 2 which can be written outside as the coefficient.

$2\sqrt{7 \times 3}$

$2\sqrt{21}$

1.193 $\sqrt{26}$

$\sqrt{13 \times 2}$

Neither 13 nor 2 is a perfect square, so $\sqrt{26}$ can't be written with a smaller radicand.

1.194 $4\sqrt{12} - 11\sqrt{3} + 2\sqrt{75}$

The radicands aren't identical. Simplify them first.

$4\sqrt{4 \times 3} - 11\sqrt{3} + 2\sqrt{25 \times 3}$

$4 \times 2\sqrt{3} - 11\sqrt{3} + 2 \times 5\sqrt{3}$

$8\sqrt{3} - 11\sqrt{3} + 10\sqrt{3}$

$(8 - 11 + 10)\sqrt{3}$

$7\sqrt{3}$

1.195 $-3\sqrt{72} \times (-2)\sqrt{5}$

Multiply the coefficients, and combine the radicands under one radical sign.

$-3 \times (-2)\sqrt{72 \times 5}$

$6\sqrt{36 \times 2 \times 5}$

$6 \times 6\sqrt{2 \times 5}$

$36\sqrt{10}$

1.196 $-6\sqrt{44} \div (-5)\sqrt{18}$

$\dfrac{-6\sqrt{44}}{-5\sqrt{18}}$

Write as a product of two separate fractions: one containing the coefficients and the other containing the radicals.

$\dfrac{-6}{-5} \times \dfrac{\sqrt{44}}{\sqrt{18}}$

Combine the radicands under one radical sign.

$\dfrac{-6}{-5} \times \sqrt{\dfrac{44}{18}}$

$\dfrac{6}{5} \times \sqrt{\dfrac{22}{9}}$

Break the radical into two separate radicals.

$\dfrac{6}{5} \times \dfrac{\sqrt{22}}{\sqrt{9}}$

$\dfrac{\cancel{6}^2}{5} \times \dfrac{\sqrt{22}}{\cancel{3}^1}$

$\dfrac{2 \times \sqrt{22}}{5 \times 1}$

$\dfrac{2\sqrt{22}}{5}$

CHAPTER 2: POLYNOMIALS & EXPRESSIONS

2.005 $13(2) + 6(2)(-1) - 1$

$26 + (-12) - 1$

$26 - 12 - 1$

13

2.006 $\dfrac{4 \cdot (0) + 9 \cdot (8)}{0 - 8}$

$\dfrac{0 + 72}{-8}$

$\dfrac{72}{-8}$

-9

2.007 $2^3 + \dfrac{4 \times 3 \times 1^3}{3}$

$8 + \dfrac{4 \times 3 \times 1}{3}$

$8 + \dfrac{12}{3}$

$8 + 4$

12

2.008 $(-3)^3 + 2(-3)^2 - (-3) + 7$

$-27 + 2(9) + 3 + 7$

$-27 + 18 + 10$

1

2.016 $(-a^2+3a-1)+(8a^2+5a-2)$
Group the *like terms*.
$$\overbrace{-a^2+8a^2} \quad \overbrace{+3a+5a} \quad \overbrace{-1+(-2)}$$
$$-a^2+8a^2 \quad\quad +3a+5a \quad\quad -1-2$$
$$7a^2+8a-3$$

2.017 $(c^2y+5)-(-2yc^2+12k)+(6-6k)$
Group the *like terms*.
$$\overbrace{c^2y-(-2yc^2)} \quad \overbrace{+5+6} \quad \overbrace{-12k+(-6k)}$$
Eliminate the excess signs.
$$\overbrace{c^2y+2yc^2} \quad \overbrace{+5+6} \quad \overbrace{-12k-6k}$$
$$3c^2y+11-18k$$

2.018
$(7ab+10-b)-(17-2ba+c)-(4ab-6c)$
Group the *like terms*.
$$\overbrace{7ab-(-2ba)-4ab} \quad \overbrace{+10-17}$$
$$\overbrace{-b} \quad \overbrace{-c-(-6c)}$$
Eliminate the excess signs.
$$\overbrace{7ab+2ba-4ab} \quad \overbrace{+10-17} \quad \overbrace{-b} \quad \overbrace{-c+6c}$$
$$5ab-7-b+5c$$

2.019 $(3p+4q)-(2p+8q)+(5p-6)$
Group the *like terms*.
$$\overbrace{3p-2p+5p} \quad \overbrace{+4q-8q} \quad +(-6)$$
Eliminate the excess signs.
$$\overbrace{3p-2p+5p} \quad \overbrace{+4q-8q} \quad -6$$
$$6p-4q-6$$

2.025 $4(x+3y)$
Distribute 4 over each inner term.
$$\overbrace{4\cdot(x)} \quad \overbrace{4\cdot(3y)}$$
$$4x+12y$$

2.026 $-2(5a+3a^2b)$
Distribute -2 onto each interior term.
$$\overbrace{-2(5a)} \quad \overbrace{-2(3a^2b)}$$
$$-10a-6a^2b$$

2.027 $ab(-a+b^2-2ab)$
Distribute ab over each interior term.
$$\overbrace{ab(-a)} \quad \overbrace{+ab(b^2)} \quad \overbrace{+ab(-2ab)}$$
$$-a^2b+ab^3-2a^2b^2$$

2.028 $3xy(2x-4y+6)$
Distribute $3xy$ onto each inner term.
$$\overbrace{3xy(2x)} \quad \overbrace{+3xy(-4y)} \quad \overbrace{+3xy(6)}$$
$$6x^2y-12xy^2+18xy$$

2.029 $-p^2q(11p-12pq^2)$
Distribute $-p^2q$ over each inner term.
$$\overbrace{-p^2q(11p)} \quad \overbrace{-p^2q(-12pq^2)}$$
$$-11p^3q+12p^3q^3$$

2.030 $4mn(3n^2+5m-mn^2)$
Distribute $4mn$ onto each interior term.
$$\overbrace{4mn(3n^2)} \quad \overbrace{+4mn(5m)} \quad \overbrace{+4mn(-mn^2)}$$
$$12mn^3+20m^2n-4m^2n^3$$

2.035 $(c+6)(c-1)$
Distribute c onto $(c-1)$. Then distribute 6 onto $(c-1)$.
$$\overbrace{c\cdot c} \quad \overbrace{+c\cdot(-1)} \quad \overbrace{+6\cdot c} \quad \overbrace{+6\cdot(-1)}$$
$$c^2-1c+6c-6$$
Combine the *like terms* $-1c$ and $6c$.
$$c^2+5c-6$$

2.036 $(4x-3y)(x+2y)$
Distribute $4x$ onto $(x+2y)$. Then distribute $-3y$ onto $(x+2y)$.
$$\overbrace{4x\cdot x} \quad \overbrace{+4x\cdot 2y} \quad \overbrace{-3y\cdot x} \quad \overbrace{-3y\cdot 2y}$$
$$4x^2+8xy-3yx-6y^2$$
Combine the *like terms* $8xy$ and $-3yx$.
$$4x^2+5xy-6y^2$$

2.037 $(m+2a)(3m-5)$
Distribute m over $(3m-5)$. Then distribute $2a$ over $(3m-5)$.
$$\overbrace{m(3m)} \quad \overbrace{+m(-5)} \quad \overbrace{+2a(3m)} \quad \overbrace{+2a(-5)}$$
$$3m^2-5m+6am-10a$$
There are no *like terms*. This is the final answer.

2.038 $(4+3p)(4-3p)$
Distribute 4 over $(4-3p)$. Then distribute $3p$ over $(4-3p)$.
$$\overbrace{4(4)} \quad \overbrace{+4(-3p)} \quad \overbrace{+3p(4)} \quad \overbrace{+3p(-3p)}$$
$$16-12p+12p-9p^2$$
Combine the *like terms* $-12p$ and $12p$ (which cancel each other out).
$$16-9p^2$$

2.046 $12a+12b$
Only 12 is common to both terms. Factor it out.
$$12(a+b)$$

2.047 $6xy+16xyz$

The GCF of 6 and 16 is 2. The GCF of xy and xyz is xy. Factor out $2xy$.

$2xy(3+8z)$

2.048 $25c^2+10c-5$

The GCF of 25, 10 and −5 is 5. There's no variable in the last term. Meaning, no variable can be factored out. Factor out 5.

$5(5c^2+2c-1)$

2.049 $-4x^3-10x^2-22x$

The GCF of 4, 10 and 22 is 2. The GCF of x^3, x^2 and x is just x. The (−) sign is also common to all terms. Factor out $-2x$.

$-2x(2x^2+5x+11)$

2.050 $8r^2-80r+16r^3$

The GCF of 8, −80 and 16 is 8. The GCF of r^2, r and r^3 is just r. Factor out $8r$.

$8r(r-10+2r^2)$

Optionally, write the polynomial inside the parentheses in reducing order of degrees.

$8r(2r^2+r-10)$

2.051 $-5mn+15m^2n^2-25mn^2$

The GCF of −5, 15 and −25 is 5. The GCF of mn, m^2n^2 and mn^2 is mn. Factor out $5mn$.

$5mn(-1+3mn-5n)$

Notice that when $5mn$ was factored out of the first term, it left behind −1. It didn't leave behind *negative nothing*.

2.060 d^2 means $1d^2$. Comparing $1d^2+12d+11$ with the quadratic ax^2+bx+c, we see that $a=1$, $b=12$, $c=11$.

The product ac is $1\times11=11$. The only way to obtain it is 1×11.

The coefficient of the middle term, 12, can be obtained by adding 1 and 11. Therefore, split the middle term as $1d+11d$.

$1d^2+1d+11d+11$

Group the first two terms. Group the next two terms. Factor the most you can from each group.

$\overbrace{1d^2+1d}\quad\overbrace{+11d+11}$

$1d(d+1)+11(d+1)$

Factor out $(d+1)$, leaving behind $1d$ and $+11$.

$(d+1)(1d+11)$

$(d+1)(d+11)$

2.061 j^2 means $1j^2$. Comparing $1j^2-16j+39$ with the quadratic ax^2+bx+c, we see that $a=1$, $b=-16$, $c=39$.

The product ac is $1\times39=39$. The different ways to obtain it are 1×39, $-1\times(-39)$, 3×13, $-3\times(-13)$.

The coefficient of the middle term, −16, can be obtained by adding −3 and −13. Therefore, split the middle term as $(-3j)+(-13j)$.

$1j^2+(-3j)+(-13j)+39$

Group the first two terms. Group the next two terms.

$\overbrace{1j^2+(-3j)}\quad\overbrace{+(-13j)+39}$

Simplify the excess signs between the terms. Factor the most you can from each group.

$\overbrace{1j^2-3j}\quad\overbrace{-13j+39}$

$1j(j-3)-13(j-3)$

Factor out $(j-3)$, leaving behind $1j$ and −13.

$(j-3)(1j-13)$

$(j-3)(j-13)$

2.062 Comparing $7h^2+12h+5$ with the quadratic ax^2+bx+c, note that $a=7$, $b=12$, $c=5$. The product ac is $7\times5=35$. The different ways to obtain it are 1×35, 5×7.

The coefficient of the middle term, 12, can be obtained by adding 5 and 7. Therefore, split the middle term as $5h+7h$.

$7h^2+5h+7h+5$

Group the first two terms. Group the next two terms. Factor the most you can from each group.

$\overbrace{7h^2+5h}\quad\overbrace{+7h+5}$

$h(7h+5)+1(7h+5)$

Notice that from the second set of parentheses, I factored out 1. *Being unable to factor out anything means factoring out just 1, because 1 times any term is that term itself.*

Factor out $(7h+5)$, leaving behind h and +1.

$(7h+5)(h+1)$

2.063 Comparing $9k^2-15k+4$ with the quadratic ax^2+bx+c, note that $a=9$, $b=-15$, $c=4$. The product ac is $9\times4=36$. The different ways to obtain it are 1×36, $-1\times(-36)$, 2×18, $-2\times(-18)$, 3×12, $-3\times(-12)$, 4×9, $-4\times(-9)$, 6×6, $-6\times(-6)$.

The coefficient of the middle term, –15, can be obtained by adding –3 and –12. Therefore, split the middle term as $(-3k)+(-12k)$.

$$9k^2+(-3k)+(-12k)+4$$

Group the first two terms. Group the next two terms.

$$\overbrace{9k^2+(-3k)}\quad\overbrace{+(-12k)+4}$$

Simplify the excess signs between the terms. Factor the most you can from each group.

$$\overbrace{9k^2-3k}\quad\overbrace{-12k+4}$$
$$3k(3k-1)-4(3k-1)$$

Factor out $(3k-1)$, leaving behind $3k$ and –4.

$$(3k-1)(3k-4)$$

2.064 r^2 means $1r^2$. Comparing $1r^2+48r-100$ with the quadratic ax^2+bx+c, we see that $a=1$, $b=48$, $c=-100$.

The product ac is $1\times(-100)=-100$. The different ways to obtain it are $1\times(-100)$, -1×100, $2\times(-50)$, -2×50, $4\times(-25)$, -4×25, $5\times(-20)$, -5×20, $10\times(-10)$.

The coefficient of the middle term, 48, can be obtained by adding –2 and 50. Therefore, split the middle term as $(-2r)+50r$.

$$1r^2+(-2r)+50r-100$$

Group the first two terms. Group the next two terms.

$$\overbrace{1r^2+(-2r)}\quad\overbrace{+50r-100}$$

Simplify the excess signs. Then factor the terms.

$$\overbrace{1r^2-2r}\quad\overbrace{+50r-100}$$
$$1r(r-2)+50(r-2)$$

Factor out $(r-2)$, leaving behind $1r$ and +50.

$$(r-2)(1r+50)$$
$$(r-2)(r+50)$$

2.065 Comparing $12t^2+5t-2$ with the quadratic ax^2+bx+c, note that $a=12$, $b=5$, $c=-2$. The product ac is $12\times(-2)=-24$. The different ways to obtain it are $1\times(-24)$, -1×24, $2\times(-12)$, -2×12, $3\times(-8)$, -3×8, $4\times(-6)$, -4×6.

The coefficient of the middle term, 5, can be obtained by adding –3 and 8. Therefore, split the middle term as $(-3t)+8t$.

$$12t^2+(-3t)+8t-2$$

Group the first two terms. Group the next two terms.

$$\overbrace{12t^2+(-3t)}\quad\overbrace{+8t-2}$$

Simplify the excess signs between the terms. Factor the most you can from each group.

$$\overbrace{12t^2-3t}\quad\overbrace{+8t-2}$$
$$3t(4t-1)+2(4t-1)$$

Factor out $(4t-1)$, leaving behind $3t$ and +2.

$$(4t-1)(3t+2)$$

2.066 u^2 means $1u^2$. Comparing $1u^2-23u-50$ with the quadratic ax^2+bx+c, we see that $a=1$, $b=-23$, $c=-50$.

The product ac is $1\times(-50)=-50$. The different ways to obtain it are $1\times(-50)$, -1×50, $2\times(-25)$, -2×25, $5\times(-10)$, -5×10.

The coefficient of the middle term, –23, can be obtained by adding 2 and –25. Therefore, split the middle term as $2u+(-25u)$.

$$1u^2+2u+(-25u)-50$$

Group the first two terms. Group the next two terms.

$$\overbrace{1u^2+2u}\quad\overbrace{+(-25u)-50}$$

Simplify the excess signs between the terms. Factor the most you can from each group.

$$\overbrace{1u^2+2u}\quad\overbrace{-25u-50}$$
$$1u(u+2)-25(u+2)$$

Factor out $(u+2)$, leaving behind $1u$ and –25.

$$(u+2)(1u-25)$$
$$(u+2)(u-25)$$

2.067 Comparing $15w^2-2w-1$ with the quadratic ax^2+bx+c, note that $a=15$, $b=-2$, $c=-1$. The product ac is $15\times(-1)=-15$. The different ways to obtain it are $1\times(-15)$, -1×15, $3\times(-5)$, -3×5.

The coefficient of the middle term, –2, can be obtained by adding 3 and –5. Therefore, split the middle term as $3w+(-5w)$.

$$15w^2+3w+(-5w)-1$$

Group the first two terms. Group the next two terms.

$$\overbrace{15w^2+3w}\quad\overbrace{+(-5w)-1}$$

Simplify the excess signs between the terms. Factor the most you can from each group.

$$\overbrace{15w^2+3w}\quad\overbrace{-5w-1}$$
$$3w(5w+1)-1(5w+1)$$

Factor out $(5w+1)$, leaving behind $3w$ and -1.

$$(5w+1)(3w-1)$$

2.068 $g^2-h^2=(g+h)(g-h)$

2.069 $L^2-R^2=(L+R)(L-R)$

2.070 $b^2-9c^2 = b^2-3^2c^2$
$$= (b+3c)(b-3c)$$

2.071 $16f^2-4j^2 = 4^2f^2-2^2j^2$
$$= (4f+2j)(4f-2j)$$

2.072 $a^2-49 = a^2-7^2 = (a+7)(a-7)$

2.073 $64y^2-x^2 = 8^2y^2-x^2$
$$= (8y+x)(8y-x)$$

2.074 $36-m^2 = 6^2-m^2 = (6+m)(6-m)$

2.075 $25x^2-1 = 5^2x^2-1^2$
$$= (5x+1)(5x-1)$$

2.076 $1-100k^2 = 1^2-10^2k^2$
$$= (1+10k)(1-10k)$$

2.085 $\dfrac{8}{x}+\dfrac{3}{x}-\dfrac{2}{x}$

Denominators are same. Add the numerators. Keep the denominator unchanged.

$$\frac{8+3-2}{x}$$

$$\frac{9}{x}$$

2.086 $\dfrac{x+7}{x-10}+\dfrac{x-2}{x-10}$

Denominators are same. Add the numerators. Keep the denominator unchanged.

$$\frac{x+7+x-2}{x-10}$$

$$\frac{2x+5}{x-10}$$

2.087 $\dfrac{2x+9}{x+7}-\dfrac{3x-4}{x+7}$

Denominators are same. Combine the numerators. Keep the denominator unchanged.

$$\frac{2x+9-(3x-4)}{x+7}$$

$$\frac{2x+9-3x+4}{x+7}$$

$$\frac{-1x+13}{x+7}$$

$$\frac{-x+13}{x+7}$$

2.088 $\dfrac{x+5}{x+1}+\dfrac{x-1}{x-2}$

Denominators are different, their LCM being $(x+1)(x-2)$. Multiply the top and bottom of the first rational expression by $(x-2)$, and the top and bottom of the second one by $(x+1)$.

$$\frac{(x+5)\cdot(x-2)}{(x+1)\cdot(x-2)}+\frac{(x-1)\cdot(x+1)}{(x-2)\cdot(x+1)}$$

Combine the numerators.

$$\frac{(x+5)\cdot(x-2)+(x-1)\cdot(x+1)}{(x-2)\cdot(x+1)}$$

Distribute and combine *like terms* in the numerator.

$$\frac{x^2-2x+5x-10+x^2+1x-1x-1}{(x-2)(x+1)}$$

$$\frac{2x^2+3x-11}{(x-2)(x+1)}$$

2.089 $\dfrac{4}{x-1}-\dfrac{x-3}{x+6}$

Denominators are different, their LCM being $(x-1)(x+6)$. Multiply the top and bottom of the first rational expression by $(x+6)$, and the top and bottom of the second one by $(x-1)$.

$$\frac{4\cdot(x+6)}{(x-1)\cdot(x+6)}-\frac{(x-3)\cdot(x-1)}{(x+6)\cdot(x-1)}$$

Combine the numerators.

$$\frac{4\cdot(x+6)-(x-3)\cdot(x-1)}{(x+6)\cdot(x-1)}$$

Distribute and combine *like terms* in the numerator. In this step, keep the $(-)$ sign outside when you distribute the second set of parentheses.

$$\frac{4x+24-(x^2-1x-3x+3)}{(x+6)(x-1)}$$

Now remove the parentheses in the numerator and distribute the $(-)$ sign onto the inner terms.

$$\frac{4x+24-x^2+1x+3x-3}{(x+6)(x-1)}$$

$$\frac{-x^2+8x+21}{(x+6)(x-1)}$$

2.090 $\dfrac{x+4}{x-3} - \dfrac{x-2}{x+3}$

Denominators are different, their LCM being $(x-3)(x+3)$. Multiply the top and bottom of the first rational expression by $(x+3)$, and the top and bottom of the second one by $(x-3)$.

$$\dfrac{(x+4)\cdot(x+3)}{(x-3)\cdot(x+3)} - \dfrac{(x-2)\cdot(x-3)}{(x+3)\cdot(x-3)}$$

Combine the numerators.

$$\dfrac{(x+4)\cdot(x+3)-(x-2)\cdot(x-3)}{(x+3)\cdot(x-3)}$$

Distribute and combine *like terms* in the numerator. In this step, keep the (–) sign outside when you distribute the second set of parentheses.

$$\dfrac{x^2+3x+4x+12-(x^2-3x-2x+6)}{(x+3)(x-3)}$$

Now remove the parentheses in the numerator and distribute the (–) sign onto the inner terms.

$$\dfrac{x^2+3x+4x+12-x^2+3x+2x-6}{(x+3)(x-3)}$$

$$\dfrac{12x+6}{(x+3)(x-3)}$$

$$\dfrac{6(2x+1)}{(x+3)(x-3)}$$

2.097 $\dfrac{x+8}{5x} \times \dfrac{x+1}{x+8}$

Cross reduce the $(x+8)$ term.

$$\dfrac{\overset{1}{\cancel{x+8}}}{5x} \times \dfrac{x+1}{\underset{1}{\cancel{x+8}}}$$

Multiply straight across.

$$\dfrac{1\cdot(x+1)}{5x\cdot1}$$

$$\dfrac{x+1}{5x}$$

2.098 $\dfrac{3x-6}{6x+6} \times \dfrac{2}{x-2}$

Factor the terms $(3x-6)$ and $(6x+6)$.

$$\dfrac{3(x-2)}{6(x+1)} \times \dfrac{2}{x-2}$$

$$\dfrac{3\overset{1}{\cancel{(x-2)}}}{\underset{3}{\cancel{6}}(x+1)} \times \dfrac{\overset{1}{\cancel{2}}}{\underset{1}{\cancel{x-2}}}$$

$$\dfrac{3\cdot1\cdot1}{3\cdot(x+1)\cdot1}$$

$$\dfrac{\overset{1}{\cancel{3}}}{\underset{1}{\cancel{3}}(x+1)}$$

$$\dfrac{1}{x+1}$$

2.099 $\dfrac{2x-16}{x^2-64} \times \dfrac{x+8}{10x}$

The factorization of $2x-16$ is $2(x-8)$.
The factorization of x^2-64 is:

$$x^2-64 = x^2-8^2 = (x+8)(x-8)$$

The original expression now becomes:

$$\dfrac{2(x-8)}{(x+8)(x-8)} \times \dfrac{x+8}{10x}$$

$$\dfrac{\overset{1}{\cancel{2(x-8)}}}{\underset{1}{\cancel{(x+8)(x-8)}}} \times \dfrac{\overset{1}{\cancel{x+8}}}{\underset{5}{\cancel{10x}}}$$

$$\dfrac{1\cdot1\cdot1}{1\cdot1\cdot5x}$$

$$\dfrac{1}{5x}$$

2.100 $\dfrac{x^2-1}{x^2-x-2} \times \dfrac{4x-8}{12}$

The factorization of $4x-8$ is $4(x-2)$.
The factorization of x^2-1 is:

$$x^2-1 = x^2-1^2 = (x+1)(x-1)$$

The factorization of x^2-x-2 is:

$$x^2-1x-2$$
$$x^2-2x+1x-2$$
$$x(x-2)+1(x-2)$$
$$(x-2)(x+1)$$

The original expression now becomes:

$$\dfrac{\overset{1}{\cancel{(x+1)}}(x-1)}{\underset{1}{\cancel{(x-2)}}\underset{1}{\cancel{(x+1)}}} \times \dfrac{\overset{1}{\cancel{4}}\overset{1}{\cancel{(x-2)}}}{\underset{3}{\cancel{12}}}$$

$$\dfrac{1\cdot(x-1)\cdot1\cdot1}{1\cdot1\cdot3}$$

$$\dfrac{x-1}{3}$$

2.101 $\dfrac{5}{2(x-2)} \times \dfrac{x^2-5x+6}{x^2-9}$

The factorization of x^2-9 is $(x+3)(x-3)$.
The factorization of x^2-5x+6 is:

$$x^2-3x-2x+6$$
$$x(x-3)-2(x-3)$$
$$(x-3)(x-2)$$

The original expression now becomes:

$$\frac{5}{2\cancel{(x-2)}^{\,1}}\times\frac{^1\cancel{(x-3)}\cancel{(x-2)}^{\,1}}{(x+3)\cancel{(x-3)}_1}$$

$$\frac{5\cdot1\cdot1}{2\cdot1\cdot(x+3)\cdot1}$$

$$\frac{5}{2(x+3)}$$

2.102 $\dfrac{2}{2x-18}\times\dfrac{x^2-81}{x+9}$

The factorization of $2x-18$ is $2(x-9)$.

The factorization of x^2-81 is:

$$x^2-81 \;=\; x^2-9^2 \;=\; (x+9)(x-9)$$

The original expression now becomes:

$$\frac{\cancel{2}^{\,1}}{_1\,2\cancel{(x-9)}_1}\times\frac{^1\cancel{(x+9)}\cancel{(x-9)}^{\,1}}{\cancel{x+9}_1}$$

$$\frac{1\cdot1\cdot1}{1\cdot1\cdot1}$$

$$1$$

2.106 $\dfrac{x-1}{x^2-2x}\div\dfrac{x^2-1}{(x-2)(x+1)}$

Multiply the first rational expression by the reciprocal of the second.

$$\frac{x-1}{x^2-2x}\times\frac{(x-2)(x+1)}{x^2-1}$$

Factor x^2-2x and x^2-1.

$$\frac{x-1}{x(x-2)}\times\frac{(x-2)(x+1)}{(x+1)(x-1)}$$

$$\frac{^1\cancel{x-1}}{x\cancel{(x-2)}_1}\times\frac{^1\cancel{(x-2)}\cancel{(x+1)}^{\,1}}{_1\cancel{(x+1)}\cancel{(x-1)}_1}$$

$$\frac{1\cdot1\cdot1}{x\cdot1\cdot1\cdot1}$$

$$\frac{1}{x}$$

2.107 $\dfrac{3x+12}{x^2+9x+20}\div\dfrac{x-2}{x^2-25}$

Multiply the first rational expression by the reciprocal of the second.

$$\frac{3x+12}{x^2+9x+20}\times\frac{x^2-25}{x-2}$$

Factorization of $3x+12$ is $3(x+4)$.

Factorization of x^2-25 is $(x+5)(x-5)$.

Factorization of $x^2+9x+20$ is:

$$1x^2+9x+20$$
$$1x^2+4x+5x+20$$
$$1x(x+4)+5(x+4)$$
$$(x+4)(1x+5)$$
$$(x+4)(x+5)$$

The original expression now becomes:

$$\frac{3(x+4)}{(x+4)(x+5)}\times\frac{(x+5)(x-5)}{x-2}$$

$$\frac{3\cancel{(x+4)}^{\,1}}{_1\cancel{(x+4)}\cancel{(x+5)}_1}\times\frac{^1\cancel{(x+5)}(x-5)}{x-2}$$

$$\frac{3\cdot1\cdot1\cdot(x-5)}{1\cdot1\cdot(x-2)}$$

$$\frac{3(x-5)}{x-2}$$

2.108 $\dfrac{x^2-36}{x-4}\div\dfrac{2x-12}{2x-8}$

Multiply the first rational expression by the reciprocal of the second.

$$\frac{x^2-36}{x-4}\times\frac{2x-8}{2x-12}$$

The factorization of x^2-36 is $(x+6)(x-6)$.

The factorization of $2x-8$ is $2(x-4)$.

The factorization of $2x-12$ is $2(x-6)$.

The original expression now becomes:

$$\frac{(x+6)(x-6)}{x-4}\times\frac{2(x-4)}{2(x-6)}$$

$$\frac{(x+6)\cancel{(x-6)}^{\,1}}{_1\cancel{x-4}}\times\frac{^1\cancel{2}\cancel{(x-4)}^{\,1}}{_1\cancel{2}\cancel{(x-6)}_1}$$

$$\frac{(x+6)\cdot1\cdot1\cdot1}{1\cdot1\cdot1}$$

$$\frac{x+6}{1}$$

$$x+6$$

2.119 $\dfrac{x^2\cdot x^7}{p^6\cdot p^3}$

$$\dfrac{x^{2+7}}{p^{6+3}}\quad\text{Refer Law 1 (page 43).}$$

$$\dfrac{x^9}{p^9}$$

$$\left(\dfrac{x}{p}\right)^9\quad\text{Refer Law 8.}$$

2.120 $(2\,m)^3 \div (4\,m)^2$

$\dfrac{(2\,m)^3}{(4\,m)^2}$

$\dfrac{2^3\,m^3}{4^2\,m^2}$ Refer Law 7 (page 43).

$\dfrac{8\,m^3}{16\,m^2}$

$\dfrac{m^3}{2\,m^2}$

$\dfrac{m^{3-2}}{2}$ Refer Law 2.

$\dfrac{m^1}{2}$

$\dfrac{m}{2}$ because m^1 is just m.

2.121 $p^4 \times q^4 \times r^{-4}$

$\dfrac{p^4 \times q^4}{r^4}$ Refer Law 4 (page 43).

$\dfrac{(p \cdot q)^4}{r^4}$ Refer Law 7.

$\left(\dfrac{pq}{r}\right)^4$ Refer Law 8.

2.122 $(7^{-2})^3 \times 7^8$

$7^{-2 \times 3} \times 7^8$ Refer Law 3 (page 43).

$7^{-6} \times 7^8$

7^{-6+8} Refer Law 1.

7^2

7×7

49

2.123 $y^{-6} \times y^4$

y^{-6+4} Refer Law 1 (page 43).

y^{-2}

$\dfrac{1}{y^2}$ Refer Law 4.

2.124 $(c^8)^8 \div (c^2)^{32}$

$\dfrac{(c^8)^8}{(c^2)^{32}}$

$\dfrac{c^{8 \times 8}}{c^{2 \times 32}}$ Refer Law 3 (page 43).

$\dfrac{c^{64}}{c^{64}}$

c^{64-64} Refer Law 2.

c^0

1 Refer Law 6.

2.125 $2^{10} \times 2^{-6} \div 4^4$

$\dfrac{2^{10} \times 2^{-6}}{4^4}$

$\dfrac{2^{10+(-6)}}{4^4}$ Refer Law 1 (page 43).

$\dfrac{2^4}{4^4}$

$\left(\dfrac{2}{4}\right)^4$ Refer Law 8.

$\left(\dfrac{1}{2}\right)^4$

$\dfrac{1^4}{2^4}$ Refer Law 8.

$\dfrac{1}{16}$

2.126 $5^{-5} \div 5^{-2}$

$\dfrac{5^{-5}}{5^{-2}}$

$5^{-5-(-2)}$ Refer Law 2 (page 43).

5^{-5+2}

5^{-3}

$\dfrac{1}{5^3}$ Refer Law 4.

$\dfrac{1}{125}$

2.127 $6(-1)^2 + \dfrac{3(20)}{5(-4)}$

$6(1) + \dfrac{60}{-20}$

$6 - 3$

3

2.128 $(10x^2-4)+(7x-3x^2)+(5-8x)$

Group the *like terms*.

$$\overbrace{10x^2+(-3x^2)} \quad \overbrace{+7x+(-8x)} \quad \overbrace{-4+5}$$
$$10x^2-3x^2 \quad +7x-8x \quad -4+5$$
$$7x^2-1x+1$$
$$7x^2-x+1$$

2.129 $(-13k+14ab-9)-(2-15k+6ab)$

Group the *like terms*.

$$\overbrace{-13k-(-15k)} \quad \overbrace{+14ab-6ab} \quad \overbrace{-9-2}$$
$$-13k+15k \quad +14ab-6ab \quad -9-2$$
$$2k+8ab-11$$

2.130 $-4a(3b^2+7ab-2a^2)$

$$\overbrace{-4a(3b^2)} \quad \overbrace{-4a(7ab)} \quad \overbrace{-4a(-2a^2)}$$
$$-12ab^2-28a^2b+8a^3$$

2.131 $(2y-9)(5-6y)$

Distribute $2y$ onto $(5-6y)$. Then distribute -9 onto $(5-6y)$.

$$\overbrace{2y\cdot5} \quad \overbrace{+2y\cdot(-6y)} \quad \overbrace{-9\cdot(5)} \quad \overbrace{-9\cdot(-6y)}$$
$$10y-12y^2-45+54y$$

Combine the *like terms* $10y$ and $54y$.

$$-12y^2+64y-45$$

2.132 $(t+11u)(3t-u)$

Distribute t onto $(3t-u)$. Then distribute $11u$ onto $(3t-u)$.

$$\overbrace{t\cdot3t} \quad \overbrace{+t\cdot(-u)} \quad \overbrace{+11u\cdot3t} \quad \overbrace{+11u\cdot(-u)}$$
$$3t^2-tu+33ut-11u^2$$

Combine the *like terms* $-tu$ and $33ut$.

$$3t^2+32ut-11u^2$$

2.133 $8y^2-18xy$

The GCF of 8 and -18 is 2. The GCF of y^2 and xy is y. Factor out $2y$.

$$2y(4y-9x)$$

2.134 $4pq-2qp^2+50pq^3$

The GCF of 4, -2 and 50 is 2. The GCF of pq, qp^2 and pq^3 is pq. Factor out $2pq$.

$$2pq(2-p+25q^2)$$

2.135 x^2 means $1x^2$. Comparing $1x^2-17x+30$ with the quadratic ax^2+bx+c, we see that $a=1$, $b=-17$, $c=30$.

The product ac is $1\times30=30$. The different ways to obtain it are 1×30, $-1\times(-30)$, 2×15, $-2\times(-15)$, 3×10, $-3\times(-10)$, 5×6, $-5\times(-6)$.

The coefficient of the middle term, -17, can be obtained by adding -2 and -15. Therefore, split the middle term as $(-2x)+(-15x)$.

$$1x^2+(-2x)+(-15x)+30$$

Group the first two terms. Group the next two terms.

$$\overbrace{1x^2+(-2x)} \quad \overbrace{+(-15x)+30}$$

Simplify the excess signs between the terms. Factor the most you can from each group.

$$\overbrace{1x^2-2x} \quad \overbrace{-15x+30}$$
$$1x(x-2)-15(x-2)$$

Factor out $(x-2)$, leaving behind $1x$ and -15.

$$(x-2)(1x-15)$$
$$(x-2)(x-15)$$

2.136 Comparing $2y^2-5y-12$ with the quadratic ax^2+bx+c, note that $a=2$, $b=-5$, $c=-12$.

The product ac is $2\times(-12)=-24$. The different ways to obtain it are $1\times(-24)$, -1×24, $2\times(-12)$, -2×12, $3\times(-8)$, -3×8, $4\times(-6)$, -4×6.

The coefficient of the middle term, -5, can be obtained by adding 3 and -8. Therefore, split the middle term as $3y+(-8y)$.

$$2y^2+3y+(-8y)-12$$

Group the first two terms. Group the next two terms.

$$\overbrace{2y^2+3y} \quad \overbrace{+(-8y)-12}$$

Simplify the excess signs between the terms. Factor the most you can from each group.

$$\overbrace{2y^2+3y} \quad \overbrace{-8y-12}$$
$$y(2y+3)-4(2y+3)$$

Factor out $(2y+3)$, leaving behind y and -4.

$$(2y+3)(y-4)$$

2.137 $k^2-81 = k^2-9^2 = (k+9)(k-9)$

2.138 $4n^2-49m^2 = 2^2n^2-7^2m^2$
$$= (2n+7m)(2n-7m)$$

2.139 $\dfrac{4x}{x-7}+\dfrac{4-x}{x-7}$

Denominators are the same. Add the numerators. Keep the denominator unchanged.

$$\frac{4x+4-x}{x-7}$$

$$\frac{3x+4}{x-7}$$

2.140 $\quad \dfrac{x-1}{x+3}-\dfrac{x+2}{x-8}$

Denominators are different, their LCM being $(x+3)(x-8)$. Multiply the top and bottom of the first rational expression by $(x-8)$, and the top and bottom of the second one by $(x+3)$.

$$\frac{(x-1)\cdot(x-8)}{(x+3)\cdot(x-8)}-\frac{(x+2)\cdot(x+3)}{(x-8)\cdot(x+3)}$$

Combine the numerators.

$$\frac{(x-1)\cdot(x-8)-(x+2)\cdot(x+3)}{(x-8)\cdot(x+3)}$$

Distribute and combine *like terms* in the numerator.

$$\frac{x^2-8x-1x+8-(x^2+3x+2x+6)}{(x-8)(x+3)}$$

$$\frac{x^2-8x-1x+8-x^2-3x-2x-6}{(x-8)(x+3)}$$

$$\frac{-14x+2}{(x-8)(x+3)}$$

$$\frac{2(-7x+1)}{(x-8)(x+3)}$$

2.141 $\quad \dfrac{x^2-25}{x^2-2x-15}\times\dfrac{3x+9}{6x}$

The factorization of $3x+9$ is $3(x+3)$.

The factorization of x^2-25 is $(x+5)(x-5)$.

The factorization of $x^2-2x-15$ is:

$$x^2-2x-15$$
$$1x^2-2x-15$$
$$1x^2-5x+3x-15$$
$$1x(x-5)+3(x-5)$$
$$(x-5)(1x+3)$$
$$(x-5)(x+3)$$

The original expression now becomes:

$$\frac{(x+5)(x-5)}{(x-5)(x+3)}\times\frac{3(x+3)}{6x}$$

$$\frac{(x+5)\cancel{(x-5)}^{1}}{{}_{1}\cancel{(x-5)}\cancel{(x+3)}_{1}}\times\frac{{}^{1}\cancel{3}\cancel{(x+3)}^{1}}{{}_{2}\cancel{6}x}$$

$$\frac{(x+5)\cdot1\cdot1\cdot1}{1\cdot1\cdot2x}$$

$$\frac{x+5}{2x}$$

2.142 $\quad \dfrac{x^2-16}{5x^2+10x}\div\dfrac{x^2+2x-8}{5x^2-10x}$

Multiply the first rational expression by the reciprocal of the second.

$$\frac{x^2-16}{5x^2+10x}\times\frac{5x^2-10x}{x^2+2x-8}$$

The factorization of x^2-16 is $(x+4)(x-4)$.

The factorization of $5x^2-10x$ is $5x(x-2)$.

The factorization of $5x^2+10x$ is $5x(x+2)$.

The factorization of x^2+2x-8 is:

$$x^2+2x-8$$
$$1x^2+2x-8$$
$$1x^2+4x-2x-8$$
$$1x(x+4)-2(x+4)$$
$$(x+4)(1x-2)$$
$$(x+4)(x-2)$$

The original expression now becomes:

$$\frac{(x+4)(x-4)}{5x(x+2)}\times\frac{5x(x-2)}{(x+4)(x-2)}$$

$$\frac{{}^{1}\cancel{(x+4)}(x-4)}{{}_{1}\cancel{5x}(x+2)}\times\frac{{}^{1}\cancel{5x}\cancel{(x-2)}^{1}}{{}_{1}\cancel{(x+4)}\cancel{(x-2)}^{1}}$$

$$\frac{1\cdot(x-4)\cdot1\cdot1}{1\cdot(x+2)\cdot1\cdot1}$$

$$\frac{x-4}{x+2}$$

2.143 $\quad \dfrac{(a^2)^4\cdot a^{-5}}{x^{-3}\cdot x^4}$

$$\frac{a^{2\times4}\cdot a^{-5}}{x^{-3}\cdot x^4}\qquad \text{Refer Law 3 (page 43).}$$

$$\frac{a^8\cdot a^{-5}}{x^{-3}\cdot x^4}$$

$$\frac{a^{8+(-5)}}{x^{-3+4}}\qquad \text{Refer Law 1.}$$

$$\frac{a^3}{x^1}\quad \text{which is just}\quad \frac{a^3}{x}.$$

2.144 $p^3 \cdot p^{-7} \cdot (q^2)^2 \cdot r^4$

$p^3 \cdot p^{-7} \cdot q^{2 \times 2} \cdot r^4$ Refer Law 3 (page 43).

$p^{3+(-7)} \cdot q^4 \cdot r^4$ Refer Law 1.

$p^{-4} \cdot q^4 \cdot r^4$

$\dfrac{q^4 \cdot r^4}{p^4}$ Refer Law 4.

$\left(\dfrac{qr}{p}\right)^4$ Refer Law 8.

CHAPTER 3: EQUATIONS

3.013 $2x - 1 = 15$

The 1 goes first; the 2 goes next.

Add 1 to both sides.

$2x - 1 = 15$
   $+1 \quad +1$

$2x = 16$

Divide both sides by 2.

$\dfrac{2x}{2} = \dfrac{16}{2}$

$x = 8$

3.014 $7y + 5 = 26$

The 5 goes first; the 7 goes next.

Subtract 5 from both sides.

$7y + 5 = 26$
   $-5 \quad -5$

$7y = 21$

Divide both sides by 7.

$\dfrac{7y}{7} = \dfrac{21}{7}$

$y = 3$

3.015 $-a + 8 = 3a - 12$

Eliminate $3a$ from the right by subtracting it from both sides.

$-a + 8 = 3a - 12$
   $-3a \qquad -3a$

$-4a + 8 = -12$

Subtract 8 from both sides.

$-4a + 8 = -12$
   $-8 \qquad -8$

$-4a = -20$

Divide both sides by –4.

$\dfrac{-4a}{-4} = \dfrac{-20}{-4}$

$a = 5$

3.016 $\dfrac{2}{5}x + 7 = 4 - \dfrac{3}{5}x$

Eliminate the x term from the right by adding $\dfrac{3}{5}x$ to both sides.

$\dfrac{2}{5}x + 7 = 4 - \dfrac{3}{5}x$
$+\frac{3}{5}x \qquad\qquad +\frac{3}{5}x$

$\dfrac{2}{5}x + \dfrac{3}{5}x + 7 = 4$

$\dfrac{5}{5}x + 7 = 4$

$1x + 7 = 4$

$x + 7 = 4$

Subtract 7 from both sides.

$x + 7 = 4$
   $-7 \quad -7$

$x = -3$

3.017 $15 + b = 5(b - 8)$

Distribute 5 onto the inner terms.

$15 + b = 5b - 40$

Eliminate $5b$ from the right by subtracting it from both sides.

$15 + b = 5b - 40$
   $-5b \qquad -5b$

$15 - 4b = -40$

Eliminate 15 from the left by subtracting it from both sides.

$15 - 4b = -40$
$-15 \qquad\qquad -15$

$-4b = -55$

Divide both sides by –4.

$\dfrac{-4b}{-4} = \dfrac{-55}{-4}$

$b = \dfrac{55}{4} = 13\dfrac{3}{4}$

3.018 $-3(m - 8) = -4(3 - 5m)$

First carry out the distribution on both sides.

$-3m + 24 = -12 + 20m$

Eliminate $20m$ from the right by subtracting it from both sides.

$-3m + 24 = -12 + 20m$
$-20m \qquad\qquad -20m$

$-23m + 24 = -12$

Subtract 24 from both sides.

$$-23m+24=-12$$
$${\scriptstyle -24}{\scriptstyle -24}$$
$$-23m=-36$$

Divide both sides by –23.

$$\frac{-23m}{-23}=\frac{-36}{-23}$$

$$m=\frac{36}{23}=1\frac{13}{23}$$

3.019 $2(x+5)-3=8$

Distribute 2 onto inner terms.

$$2x+10-3=8$$
$$2x+7=8$$

Subtract 7 from both sides.

$$2x+7=8$$
$${\scriptstyle -7}{\scriptstyle -7}$$
$$2x=1$$

Divide both sides by 2.

$$\frac{2x}{2}=\frac{1}{2}$$

$$x=\frac{1}{2}$$

3.020 $\frac{4}{7}(x+21)=6$

Divide both sides by $\frac{4}{7}$. (If you want, you can distribute first and take that route instead.)

$$\frac{\frac{4}{7}(x+21)}{\frac{4}{7}}=\frac{6}{\frac{4}{7}}$$

$$x+21=6\times\frac{7}{4}$$

$$x+21=\frac{42}{4}$$

$$x+21=\frac{21}{2}$$

$$x+21=10\frac{1}{2}$$

Subtract 21 from both sides.

$$x+21=10\frac{1}{2}$$
$${\scriptstyle -21}{\scriptstyle -21}$$
$$x=-10\frac{1}{2}$$

3.021 $3+\frac{2}{9}(c-1)=11$

3 goes first; $\frac{2}{9}$ goes next.

Subtract 3 from both sides.

$$3+\frac{2}{9}(c-1)=11$$
$${\scriptstyle -3}\phantom{+\frac{2}{9}(c-1)=}{\scriptstyle -3}$$

$$\frac{2}{9}(c-1)=8$$

Divide both sides by $\frac{2}{9}$.

$$\frac{\frac{2}{9}(c-1)}{\frac{2}{9}}=\frac{8}{\frac{2}{9}}$$

$$c-1=8\times\frac{9}{2}$$

$$c-1=\frac{72}{2}$$

$$c-1=36$$

Add 1 to both sides.

$$c-1=36$$
$${\scriptstyle +1}{\scriptstyle +1}$$
$$c=37$$

3.022 $\frac{10}{3x}+4=\frac{16}{3x}$

Eliminate the x term from the left by subtracting it from both sides. It will involve fewer steps.

$$\frac{10}{3x}+4=\frac{16}{3x}$$
$${\scriptstyle -\frac{10}{3x}}{\scriptstyle -\frac{10}{3x}}$$

$$4=\frac{16}{3x}-\frac{10}{3x}$$

$$4=\frac{6}{3x}$$

$$4=\frac{2}{x}$$

$$\frac{4}{1}=\frac{2}{x}$$

Take reciprocals of both sides.

$$\frac{1}{4}=\frac{x}{2}$$

Multiply both sides by 2.

$$\frac{1}{4}\times 2=\frac{x}{2}\times 2$$

$$\frac{2}{4}=\frac{2x}{2}$$

$$\frac{1}{2}=x$$

3.023 $\quad \frac{1}{x}+\frac{7}{x}=3-\frac{2}{x}$

Combine *like terms* on the left.

$$\frac{8}{x}=3-\frac{2}{x}$$

Eliminate the x term from the right by adding $\frac{2}{x}$ to both sides.

$$\frac{8}{x}=3-\frac{2}{x}$$
$$\underset{+\frac{2}{x}}{}\quad \underset{+\frac{2}{x}}{}$$

$$\frac{10}{x}=3$$

$$\frac{10}{x}=\frac{3}{1}$$

Take reciprocals of both sides.

$$\frac{x}{10}=\frac{1}{3}$$

Multiply both sides by 10.

$$\frac{x}{10}\times 10=\frac{1}{3}\times 10$$

$$\frac{10x}{10}=\frac{10}{3}$$

$$x=\frac{10}{3}=3\frac{1}{3}$$

3.030 Assume the number as t. Four times the number would be $4t$. Thirteen less than that would be $4t-13$. This is equal to 15.

The equation would be:

$$4t-13=15$$
$$4t-13=15$$
$$\underset{+13}{}\quad \underset{+13}{}$$
$$4t=28$$
$$\frac{4t}{4}=\frac{28}{4}$$
$$t=7$$

3.031 Assume the number as u. The sum of the number and 18 would be $u+18$. Three less than 29 is $29-3$ which is 26. These two things are apparently equal. The equation would be:

$$u+18=26$$
$$u+18=26$$
$$\underset{-18}{}\quad \underset{-18}{}$$
$$u=8$$

3.032 Assume the number as v. When it's tripled, it would be $3v$. Eighteen more than its original self would be $18+v$. These two are apparently equal. The equation would be:

$$3v=18+v$$
$$3v=18+v$$
$$\underset{-v}{}\quad \underset{-v}{}$$
$$2v=18$$
$$\frac{2v}{2}=\frac{18}{2}$$
$$v=9$$

3.033 Assume the number as w. The difference of the number and 12 would be $w-12$. Three more than twice the number would be $2w+3$. These two are apparently equal.

The equation would be:

$$w-12=2w+3$$
$$w-12=2w+3$$
$$\underset{-2w}{}\quad \underset{-2w}{}$$
$$-1w-12=3$$
$$-1w-12=3$$
$$\underset{+12}{}\quad \underset{+12}{}$$
$$-1w=15$$
$$\frac{-1w}{-1}=\frac{15}{-1}$$
$$w=-15$$

3.034 Assume the number as x. The product of the number and 5 would be $5x$. The sum of the number and 5 would be $x+5$. These two are apparently the same. The equation would be:

$$5x=x+5$$
$$5x=x+5$$
$$\underset{-x}{}\quad \underset{-x}{}$$
$$4x=5$$
$$\frac{4x}{4}=\frac{5}{4}$$
$$x=\frac{5}{4}=1\frac{1}{4}$$

3.041 $\quad \frac{m}{3}=\frac{7}{5}$

Multiply both sides by 3.

$$\frac{m}{3} \times 3 = \frac{7}{5} \times 3$$

$$\frac{30\,m}{3} = \frac{7 \times 3}{5}$$

$$m = \frac{21}{5}$$

3.042 $\quad \dfrac{10}{3} = \dfrac{5\,d}{6}$

$$\frac{10}{3} = \frac{5}{6}\,d$$

Divide both sides by $\dfrac{5}{6}$.

$$\frac{\frac{10}{3}}{\frac{5}{6}} = \frac{\frac{5}{6}\,d}{\frac{5}{6}}$$

$$\frac{^{2}\!\!\!\not{10}}{_{1}\not{3}} \times \frac{\not{6}^{\,2}}{\not{5}_{\,1}} = d$$

$$\frac{2 \times 2}{1 \times 1} = d$$

$$4 = d$$

3.043 $\quad \dfrac{4}{9\,y} = \dfrac{3}{2}$

Take reciprocals of both sides.

$$\frac{9\,y}{4} = \frac{2}{3}$$

$$\frac{9}{4}\,y = \frac{2}{3}$$

Divide both sides by $\dfrac{9}{4}$.

$$\frac{\frac{9}{4}\,y}{\frac{9}{4}} = \frac{\frac{2}{3}}{\frac{9}{4}}$$

$$y = \frac{2}{3} \times \frac{4}{9} = \frac{2 \times 4}{3 \times 9} = \frac{8}{27}$$

3.044 Assume h as the height and w as the width. Then, $h:w=9:2$. It is given that $h=27$. The proportion would then be $27:w=9:2$.
The equation would be:

$$\frac{27}{w} = \frac{9}{2}$$

Take reciprocals of both sides.

$$\frac{w}{27} = \frac{2}{9}$$

Multiply both sides by 27.

$$\frac{w}{27} \times 27 = \frac{2}{9} \times 27$$

$$\frac{27\,w}{27} = \frac{2 \times \not{27}^{\,3}}{\not{9}^{\,1}}$$

$$w = \frac{2 \times 3}{1} = 6$$

3.045 Assume m as the number of men and w as the number of women. Then, $m:w=3:5$. It is given that $m=12$. The proportion would then be $12:w=3:5$. The equation would be:

$$\frac{12}{w} = \frac{3}{5}$$

Take reciprocals of both sides.

$$\frac{w}{12} = \frac{5}{3}$$

Multiply both sides by 12.

$$\frac{w}{12} \times 12 = \frac{5}{3} \times 12$$

$$\frac{12\,w}{12} = \frac{5 \times \not{12}^{\,4}}{\not{3}^{\,1}}$$

$$w = \frac{5 \times 4}{1} = 20$$

3.046 Assume E as English score and H as history score. Then, $E:H=8:7$. It is given that $E=96$. The proportion would then be $96:H=8:7$. The equation would be:

$$\frac{96}{H} = \frac{8}{7}$$

Take reciprocals of both sides.

$$\frac{H}{96} = \frac{7}{8}$$

Multiply both sides by 96.

$$\frac{H}{96} \times 96 = \frac{7}{8} \times 96$$

$$\frac{96\,H}{96} = \frac{7 \times \not{96}^{\,12}}{\not{8}^{\,1}}$$

$$H = \frac{7 \times 12}{1} = 84$$

3.052 Solve $-7\,x + 3\,y = 5\,b$ for b.
Divide both sides by 5.

$$\frac{-7x+3y}{5}=\frac{5b}{5}$$

$$\frac{-7x+3y}{5}=b$$

3.053 Solve $v=u+at$ for t.

To solve for t, the u and a must go from the right. u goes first; a goes next. Subtract u from both sides.

$$v=u+at$$
$$_{-u}_{-u}$$
$$v-u=at$$

Divide both sides by a.

$$\frac{v-u}{a}=\frac{at}{a}$$

$$\frac{v-u}{a}=t$$

3.054 Solve $2(x+y)=18$ for x.

$2(x+y)$ means $2\times(x+y)$. Divide both sides by 2 to eliminate it from the left.

$$\frac{2(x+y)}{2}=\frac{18}{2}$$

$$x+y=9$$

Subtract y from both sides.

$$x+y=9$$
$$_{-y}_{-y}$$
$$x=9-y$$

3.055 Solve $n-ab=3ab+m$ for a.

Notice that there are a terms on both sides. Subtract $3ab$ from both sides to eliminate it from the right.

$$n-ab=3ab+m$$
$$_{-3ab}_{-3ab}$$
$$n-4ab=m$$

To solve for a, the n, 4 and b must go. n goes first; 4 and b (together) go next. Subtract n from both sides.

$$n-4ab=m$$
$$_{-n}_{-n}$$
$$-4ab=m-n$$

Divide both sides by $-4b$.

$$\frac{-4ab}{-4b}=\frac{m-n}{-4b}$$

$$a=\frac{m-n}{-4b}$$

3.056 Solve $3p+5q=mp-q$ for p.

There's a p term on both sides. Subtract mp from both sides to eliminate it from the right.

$$3p+5q=mp-q$$
$$_{-mp}_{-mp}$$
$$3p-mp+5q=-q$$

Subtract $5q$ from both sides.

$$3p-mp+5q=-q$$
$$_{-5q}_{-5q}$$
$$3p-mp=-6q$$

Factor out p on the left.

$$p(3-m)=-6q$$

Divide both sides by $(3-m)$.

$$\frac{p(3-m)}{3-m}=\frac{-6q}{3-m}$$

$$p=\frac{-6q}{3-m}$$

3.057 Solve $\frac{F}{m}=a+g$ for m.

$$\frac{F}{m}=\frac{a+g}{1}$$

To bring m to the numerator, take reciprocals of both sides.

$$\frac{m}{F}=\frac{1}{a+g}$$

Multiply both sides by F.

$$\frac{m}{F}\times F=\frac{1}{a+g}\times F$$

$$\frac{mF}{F}=\frac{F}{a+g}$$

$$m=\frac{F}{a+g}$$

3.063 $x+y=0$ and $x=7-2y$

In the second equation, x is expressed in the form of other terms. Substitute those terms for x in the first equation.

$$x+y=0$$
$$(7-2y)+y=0$$
$$7-y=0$$

$$7-y=0$$
$$_{-7}_{-7}$$
$$-y=-7$$

Multiply both sides by -1 to reverse the signs.

$$y=7$$

Now that y is known, plug it in the second equation to find the value of x.

$x=7-2y$

$x=7-2(7)$

$x=-7$

3.064 $y=9+3x$ and $2x-y=-8$

In the first equation, y is expressed in the form of other terms. Substitute those terms for y in the second equation.

$2x-y=-8$

$2x-(9+3x)=-8$

$2x-9-3x=-8$

$-x-9=-8$

$-x-9=-8$
$\quad\;\;+9\quad\;\;+9$

$-x=1$

Multiply both sides by -1 to reverse the signs.

$x=-1$

Now that x is known, plug it in the first equation to find the value of y.

$y=9+3x$

$y=9+3(-1)$

$y=6$

3.065 $-4y+2x=20$ and $x=5y+1$

In the second equation, x is expressed in the form of other terms. Substitute those terms for x in the first equation.

$-4y+2x=20$

$-4y+2(5y+1)=20$

$-4y+10y+2=20$

$6y+2=20$

$6y+2=20$
$\quad\;\;-2\quad\;\;-2$

$6y=18$

$\dfrac{6y}{6}=\dfrac{18}{6}$

$y=3$

Now that y is known, plug it in the second equation to find the value of x.

$x=5y+1$

$x=5(3)+1$

$x=16$

3.066 $y=-x-7$ and $5x=2y$

In the first equation, y is expressed in the form of other terms. Substitute those terms for y in the second equation.

$5x=2y$

$5x=2(-x-7)$

$5x=-2x-14$

$5x=-2x-14$
$+2x\quad\;\;+2x$

$7x=-14$

$\dfrac{7x}{7}=\dfrac{-14}{7}$

$x=-2$

Now that x is known, plug it in the first equation to find the value of y.

$y=-x-7$

$y=-(-2)-7$

$y=2-7$

$y=-5$

3.067 Assume n as the price of a notebook, and p as the price of a pen. Then, 3 pens would cost $3p$ dollars. From the first sentence of the problem, $n=3p-5$. From the second sentence, 6 notebooks would cost $6n$ dollars and 6 pens would cost $6p$ dollars. That total cost is given as $18. So the second equation would be $6n+6p=18$.

In the first equation, n is expressed in the form of other terms. Substitute those terms for n in the second equation.

$6n+6p=18$

$6(3p-5)+6p=18$

$18p-30+6p=18$

$24p-30=18$

$24p-30=18$
$\quad\quad\;\;+30\quad\;\;+30$

$24p=48$

$\dfrac{24p}{24}=\dfrac{48}{24}$

$p=2$

Now that p is known, plug it in the first equation to find the value of n.

$n=3p-5$

$n=3(2)-5$

$n=1$

A pen costs $2, and a notebook costs $1

3.068 Assume a glass bowl weighs G ounces, and a plastic bowl weighs P ounces. From the first sentence of the problem, $G = 3P$. From the second sentence, 5 plastic bowls would weigh $5P$ ounces, and 2 glass bowls would weigh $2G$ ounces. That total weight is given as 66 ounces. So the second equation would be $5P + 2G = 66$.
In the first equation, G is expressed in the form of the other term. Substitute that term for G in the second equation.

$5P + 2G = 66$

$5P + 2(3P) = 66$

$5P + 6P = 66$

$11P = 66$

$\dfrac{11P}{11} = \dfrac{66}{11}$

$P = 6$

Now that P is known, plug it in the first equation to find the value of G.

$G = 3P$

$G = 3(6)$

$G = 18$

A plastic bowl weighs 6 ounces, and a glass bowl weighs 18 ounces.

3.074 $5x - 3y = 13$ and $x - 3y = -7$

Multiply the second equation by -1 so that $-3y$ turns into $3y$ which is the opposite of $-3y$ in the first equation. Then add the new equation to the first equation.

$-1 \times (x - 3y) = -1 \times (-7)$

$\begin{array}{r} -x + 3y = 7 \\ + \underline{5x - 3y = 13} \\ 4x \quad\quad = 20 \end{array}$

$\dfrac{4x}{4} = \dfrac{20}{4}$

$x = 5$

Now that x is known, plug it into the second equation to find the value of y.

$x - 3y = -7$

$\underset{-5}{5} - 3y = \underset{-5}{-7}$

$-3y = -12$

$\dfrac{-3y}{-3} = \dfrac{-12}{-3}$

$y = 4$

3.075 $3x + 2y = 8$ and $5x - 7y = -28$

Eliminate the x terms, their LCM being $15x$. Multiply the first equation by -5, and the second equation by 3.

$\begin{array}{c|c} -5(3x + 2y) = -5(8) & 3(5x - 7y) = 3(-28) \\ -15x - 10y = -40 & 15x - 21y = -84 \end{array}$

Now add the two equations.

$\begin{array}{r} -15x - 10y = -40 \\ + \underline{15x - 21y = -84} \\ -31y = -124 \end{array}$

$\dfrac{-31y}{-31} = \dfrac{-124}{-31}$

$y = 4$

Now that y is known, plug it into the first equation to find the value of x.

$3x + 2y = 8$

$3x + 2(4) = 8$

$3x + 8 = 8$

$\underset{-8}{3x + 8} = \underset{-8}{8}$

$3x = 0$

$\dfrac{3x}{3} = \dfrac{0}{3}$

$x = 0$

3.076 $3x - 7y = 14$ and $x - 4y = 8$

Multiply the second equation by -3 to get $-3x$ which is the opposite of $3x$ in the first equation. Then add the new equation to the first equation.

$-3(x - 4y) = -3(8)$

$\begin{array}{r} -3x + 12y = -24 \\ + \underline{3x - 7y = 14} \\ 5y = -10 \end{array}$

$\dfrac{5y}{5} = \dfrac{-10}{5}$

$y = -2$

Now that y is known, plug it into the second equation to find the value of x.

$x - 4y = 8$

$x - 4(-2) = 8$

$x + 8 = 8$

$\underset{-8}{x + 8} = \underset{-8}{8}$

$x = 0$

3.077 Assume a T-shirt costs T dollars, and a sweater costs S dollars. The first sentence of the problem means $T+S=21$. The second sentence means $4T+2S=54$.

Eliminate the S terms from the two equations by multiplying the first equation by –2 and then adding the new equation to the second equation.

$$-2(T+S)=-2(21)$$

$$\begin{array}{r} -2T-2S=-42 \\ +\ 4T+2S=\ 54 \\ \hline 2T\qquad =\ 12 \end{array}$$

$$\frac{2T}{2}=\frac{12}{2}$$

$$T=6$$

Now that T is known, plug it into the first equation to find the value of S.

$$T+S=21$$

$$6+S=21$$

$$\underset{-6\qquad -6}{6+S=21}$$

$$S=15$$

A T-shirt costs \$6, and a sweater costs \$15.

3.078 Assume a pencil is P inches long, and an eraser is E inches long. The first part of the problem means $3P+6E=30$. The next part means $5P+4E=38$.

Eliminate the P terms, their LCM being $15P$. Multiply the first equation by –5, and the second equation by 3.

$$\begin{array}{c|c} -5(3P+6E)=-5(30) & 3(5P+4E)=3(38) \\ -15P-30E=-150 & 15P+12E=114 \end{array}$$

Now add the two equations.

$$\begin{array}{r} -15P-30E=-150 \\ +\ 15P+12E=\ \ 114 \\ \hline -18E=\ -36 \end{array}$$

$$\frac{-18E}{-18}=\frac{-36}{-18}$$

$$E=2$$

Now that E is known, plug it into the first equation to find the value of P.

$$3P+6E=30$$

$$3P+6(2)=30$$

$$3P+12=30$$

$$\underset{-12\qquad -12}{3P+12=30}$$

$$3P=18$$

$$\frac{3P}{3}=\frac{18}{3}$$

$$P=6$$

A pencil is 6 inches long, and an eraser is 2 inches long.

3.083 $\quad|14+x|=5$

$$\begin{array}{c|c} 14+x=5 & -(14+x)=5 \\ \underset{-14\qquad -14}{14+x=\ 5} & -14-x=5 \\ x=-9 & \underset{+14\qquad\ +14}{-14-x=\ 5} \\ & -x=19 \\ & x=-19 \end{array}$$

3.084 $\quad|45-3x|=33$

$$\begin{array}{c|c} 45-3x=33 & -(45-3x)=33 \\ \underset{-45\qquad -45}{45-3x=33} & -45+3x=33 \\ -3x=-12 & \underset{+45\qquad\ +45}{-45+3x=33} \\ \frac{-3x}{-3}=\frac{-12}{-3} & 3x=78 \\ x=4 & \frac{3x}{3}=\frac{78}{3} \\ & x=26 \end{array}$$

3.085 $\quad4|5x-1|-1=19$

First, get the modulus by itself.

$$\underset{+1\qquad\ +1}{4|5x-1|-1=19}$$

$$4|5x-1|=20$$

$$\frac{4|5x-1|}{4}=\frac{20}{4}$$

$$|5x-1|=5$$

Modulus is positive. Equation is valid. Now solve.

$$\begin{array}{c|c} 5x-1=5 & -(5x-1)=5 \\ \underset{+1\qquad\ +1}{5x-1=5} & -5x+1=5 \\ 5x=6 & \underset{-1\qquad\ -1}{-5x+1=5} \\ x=\frac{6}{5}=1\frac{1}{5} & -5x=4 \\ & \frac{-5x}{-5}=\frac{4}{-5} \\ & x=-\frac{4}{5} \end{array}$$

3.086 $|x+6|+12=7$

First, get the modulus by itself.

$$|x+6|\underset{-12}{+12}=\underset{-12}{7}$$

$$|x+6|=-5$$

Modulus is negative. Equation has no solution.

3.087 $-13+|x-15|=9$

First, get the absolute value by itself.

$$\underset{+13}{-13}+|x-15|=\underset{+13}{9}$$

$$|x-15|=22$$

Modulus is positive. Equation can be solved.

$x-15=22$	$-(x-15)=22$
$x-15=22$	$-x+15=22$
$\underset{+15}{}\ \underset{+15}{}$	
$x=37$	$-x+15=22$
	$\underset{-15}{}\ \underset{-15}{}$
	$-x=7$
	$x=-7$

3.088 $11+2|x+10|=5$

First, get the absolute value by itself.

$$\underset{-11}{11}+2|x+10|=\underset{-11}{5}$$

$$2|x+10|=-6$$

$$\frac{2|x+10|}{2}=\frac{-6}{2}$$

$$|x+10|=-3$$

Absolute value is negative. No solution.

3.095 $x^2+14x+24=0$

$$1x^2+14x+24=0$$
$$1x^2+12x+2x+24=0$$
$$1x(x+12)+2(x+12)=0$$
$$(x+12)(1x+2)=0$$
$$(x+12)(x+2)=0$$

This is a zero product. Equate each factor to zero.

$x+12=0$	$x+2=0$
$\underset{-12}{}\ \underset{-12}{}$	$\underset{-2}{}\ \underset{-2}{}$
$x=-12$	$x=-2$

3.096 $12x^2-4x-1=0$

$$12x^2-6x+2x-1=0$$
$$6x(2x-1)+1(2x-1)=0$$
$$(2x-1)(6x+1)=0$$

This is a zero product. Equate each factor to zero.

$2x-1=0$	$6x+1=0$
$\underset{+1}{}\ \underset{+1}{}$	$\underset{-1}{}\ \underset{-1}{}$
$2x=1$	$6x=-1$
$\dfrac{2x}{2}=\dfrac{1}{2}$	$\dfrac{6x}{6}=\dfrac{-1}{6}$
$x=\dfrac{1}{2}$	$x=-\dfrac{1}{6}$

3.097 $3x^2-x-50=2x^2+40$

$$3x^2-\underset{-2x^2}{x}-50=2x^2+\underset{-2x^2}{40}$$

$$x^2-x-50=\underset{-40}{40}$$
$$\underset{-40}{}$$

$$x^2-x-90=0$$
$$1x^2-1x-90=0$$
$$1x^2-10x+9x-90=0$$
$$1x(x-10)+9(x-10)=0$$
$$(x-10)(1x+9)=0$$
$$(x-10)(x+9)=0$$

This is a zero product. Equate each factor to zero.

$x-10=0$	$x+9=0$
$\underset{+10}{}\ \underset{+10}{}$	$\underset{-9}{}\ \underset{-9}{}$
$x=10$	$x=-9$

3.098 $2x^2+19=4-11x$

$$2x^2+\underset{-4}{19}=\underset{-4}{4}-11x$$

$$2x^2+15\ \underset{+11x}{}=-11x\underset{+11x}{}$$

$$2x^2+15+11x=0$$

$$2x^2+11x+15=0$$

Now this looks like a quadratic equation.

$$2x^2+6x+5x+15=0$$
$$2x(x+3)+5(x+3)=0$$
$$(x+3)(2x+5)=0$$

This is a zero product. Equate each factor to zero.

$x+3=0$	$2x+5=0$
$\underset{-3}{}\ \underset{-3}{}$	$\underset{-5}{}\ \underset{-5}{}$
$x=-3$	$2x=-5$
	$\dfrac{2x}{2}=\dfrac{-5}{2}$
	$x=-\dfrac{5}{2}=-2\dfrac{1}{2}$

3.099 The quadratic can't be factored. Use the quadratic formula. Comparing $x^2-6x-3=0$ with $ax^2+bx+c=0$, note that $a=1$, $b=-6$, $c=-3$.

$$x=\frac{-b\pm\sqrt{b^2-4ac}}{2a}$$

$$x = \frac{-(-6) \pm \sqrt{(-6)^2 - 4(1)(-3)}}{2(1)}$$

$$x = \frac{6 \pm \sqrt{36 + 12}}{2}$$

$$x = \frac{6 \pm \sqrt{48}}{2}$$

The two answers are:

$$x = \frac{6 + \sqrt{48}}{2} \quad \text{and} \quad x = \frac{6 - \sqrt{48}}{2}$$

Rounded to two decimal digits, the answers are:
$x = 6.46$ and $x = -0.46$

3.100 The quadratic can't be factored. Use the quadratic formula. Comparing $5x^2 - 10x + 3 = 0$ with $ax^2 + bx + c = 0$, note that $a = 5$, $b = -10$, $c = 3$.

$$x = \frac{-b \pm \sqrt{b^2 - 4ac}}{2a}$$

$$x = \frac{-(-10) \pm \sqrt{(-10)^2 - 4(5)(3)}}{2(5)}$$

$$x = \frac{10 \pm \sqrt{100 - 60}}{10}$$

$$x = \frac{10 \pm \sqrt{40}}{10}$$

The two answers are:

$$x = \frac{10 + \sqrt{40}}{10} \quad \text{and} \quad x = \frac{10 - \sqrt{40}}{10}$$

Rounded to two decimal digits, the answers are:
$x = 1.63$ and $x = 0.37$

3.107 $2^{5x} = 2^{12-x}$

The bases are the same. Equate the exponents.

$$5x = 12 - x$$
$$_{+5} _{+x}$$
$$6x = 12$$
$$\frac{6x}{6} = \frac{12}{6}$$
$$x = 2$$

3.108 $4^{5x} = 2^{x+9}$

The bases are not the same. To equate them, notice that 4 can be written as 2^2.

$$(2^2)^{5x} = 2^{x+9}$$

$$2^{10x} = 2^{x+9} \qquad \text{Refer page 43, law 3.}$$

The bases are the same. Equate the exponents.

$$10x = x + 9$$
$$_{-x} _{-x}$$
$$9x = 9$$
$$\frac{9x}{9} = \frac{9}{9}$$
$$x = 1$$

3.109 $3^{x+8} = 27$

The bases are not the same. To equate them, write 27 as 3^3.

$$3^{x+8} = 3^3$$

The bases are the same. Equate the exponents.

$$x + 8 = 3$$
$$_{-8} _{-8}$$
$$x = -5$$

3.110 $25^3 = 5^{2x-2}$

The bases are not the same. To equate them, write 25 as 5^2.

$$(5^2)^3 = 5^{2x-2}$$

$$5^6 = 5^{2x-2} \qquad \text{Refer page 43, law 3.}$$

The bases are the same. Equate the exponents.

$$6 = 2x - 2$$
$$_{+2} _{+2}$$
$$8 = 2x$$
$$\frac{8}{2} = \frac{2x}{2}$$
$$4 = x$$

3.111 $16^4 = 4^{5x-2}$

The bases are not the same. To equate them, write 16 as 4^2.

$$(4^2)^4 = 4^{5x-2}$$

$$4^8 = 4^{5x-2} \qquad \text{Refer page 43, law 3.}$$

The bases are the same. Equate the exponents.

$$8 = 5x - 2$$
$$_{+2} _{+2}$$
$$10 = 5x$$
$$\frac{10}{5} = \frac{5x}{5}$$
$$2 = x$$

3.112 $7^{2x+4} = 49$

The bases are not the same. To equate them, write 49 as 7^2.

$$7^{2x+4} = 7^2$$

The bases are the same. Equate the exponents.

$$2x+4=2$$
$${\scriptstyle-4}{\scriptstyle-4}$$
$$2x=-2$$
$$\frac{2x}{2}=\frac{-2}{2}$$
$$x=-1$$

3.118 $\sqrt{x+8}=-3$

Square both sides to eliminate the radical sign.

$$\left(\sqrt{x+8}\right)^2=(-3)^2$$
$$x+8=(-3)\cdot(-3)$$
$$x+8=9$$
$${\scriptstyle-8}{\scriptstyle-8}$$
$$x=1$$

3.119 $5\sqrt{3x}=30$

First, isolate the radical.

$$\frac{5\sqrt{3x}}{5}=\frac{30}{5}$$
$$\sqrt{3x}=6$$

Square both sides to eliminate the radical sign.

$$\left(\sqrt{3x}\right)^2=6^2$$
$$3x=6\cdot6$$
$$3x=36$$
$$\frac{3x}{3}=\frac{36}{3}$$
$$x=12$$

3.120 $10+\sqrt{5x+45}=20$

First, isolate the radical.

$$10+\sqrt{5x+45}=20$$
$${\scriptstyle-10}\phantom{+\sqrt{5x+45}=20}{\scriptstyle-10}$$
$$\sqrt{5x+45}=10$$

Square both sides to eliminate the radical sign.

$$\left(\sqrt{5x+45}\right)^2=10^2$$
$$5x+45=10\cdot10$$
$$5x+45=100$$
$${\scriptstyle-45}{\scriptstyle-45}$$
$$5x=55$$
$$\frac{5x}{5}=\frac{55}{5}$$
$$x=11$$

3.121 $25+2\sqrt{80+4x}=49$

First, isolate the radical.

$$25+2\sqrt{80+4x}=49$$
$${\scriptstyle-25}\phantom{+2\sqrt{80+4x}=}{\scriptstyle-25}$$
$$2\sqrt{80+4x}=24$$
$$\frac{2\sqrt{80+4x}}{2}=\frac{24}{2}$$
$$\sqrt{80+4x}=12$$

Square both sides to eliminate the radical sign.

$$\left(\sqrt{80+4x}\right)^2=12^2$$
$$80+4x=12\cdot12$$
$$80+4x=144$$
$${\scriptstyle-80}{\scriptstyle-80}$$
$$4x=64$$
$$\frac{4x}{4}=\frac{64}{4}$$
$$x=16$$

3.122 $2(x+4)-1=11$
$$2(x+4)-1=11$$
$${\scriptstyle+1}{\scriptstyle+1}$$
$$2(x+4)=12$$
$$\frac{2(x+4)}{2}=\frac{12}{2}$$

If you'd prefer distributing first, you can do that. In the end, you'd get the same answer.

$$x+4=6$$
$${\scriptstyle-4}{\scriptstyle-4}$$
$$x=2$$

3.123 $1-\dfrac{8}{3x}=\dfrac{1}{3x}+\dfrac{2}{3}$

Get the x terms onto the left, and the independent constants onto the right.

$$1-\frac{8}{3x}=\frac{1}{3x}+\frac{2}{3}$$
$$\phantom{1-\frac{8}{3x}}{\scriptstyle-\frac{1}{3x}}{\scriptstyle-\frac{1}{3x}}$$
$$1-\frac{9}{3x}=\frac{2}{3}$$
$$1-\frac{9}{3x}=\frac{2}{3}$$
$${\scriptstyle-1}\phantom{\frac{9}{3x}=}\frac{2}{3}{\scriptstyle-1}$$
$$-\frac{9}{3x}=\frac{2}{3}-1$$

Note that 1 is the same as $\dfrac{3}{3}$.

$$-\frac{9}{3x}=\frac{2}{3}-\frac{3}{3}$$

145

$$-\frac{9}{3x}=-\frac{1}{3}$$

To bring the x term into the numerator, take reciprocals of both sides. Also, multiply both sides by -1 to flip the signs.

$$\frac{3x}{9}=\frac{3}{1}$$

$$\frac{x}{3}=3$$

$$\frac{x}{3}\times 3=3\times 3$$

$$\frac{3x}{3}=9$$

$$x=9$$

3.124 Assume x as the number. The product of the number and 7 would be $7x$. Two less than that would be $7x-2$. Also, the sum of the number and 10 would be $x+10$. These two values are given as equal. The equation would be:

$$7x-2=x+10$$

$$7x-2=x+10$$
$$\underset{-x}{}\underset{-x}{}$$

$$6x-2=10$$
$$\underset{+2}{}\underset{+2}{}$$

$$6x=12$$

$$\frac{6x}{6}=\frac{12}{6}$$

$$x=2$$

3.125 $\quad\dfrac{4}{5c}=\dfrac{2}{15}$

To bring $5c$ into the numerator, take reciprocals of both sides.

$$\frac{5c}{4}=\frac{15}{2}$$

To eliminate 4, multiply both sides by 4.

$$\frac{5c}{4}\times 4=\frac{15}{2}\times 4$$

$$5c=\frac{15\times 4^{\,2}}{2^{\,1}}$$

$$5c=\frac{15\times 2}{1}=\frac{30}{1}=30$$

$$\frac{5c}{5}=\frac{30}{5}$$

$$c=6$$

3.126 Assume C as the number of cruisers and H as the number of choppers. Then, $C:H=3:7$. It is given that $C=93$. The proportion would then be $93:H=3:7$. The equation would be

$$\frac{93}{H}=\frac{3}{7}$$

Take reciprocals of both sides.

$$\frac{H}{93}=\frac{7}{3}$$

Multiply both sides by 93.

$$\frac{H}{93}\times 93=\frac{7}{3}\times 93$$

$$\frac{93H}{93}=\frac{7\times 93^{\,31}}{3^{\,1}}$$

$$H=\frac{7\times 31}{1}=217$$

3.127 Solve $3a(b-c)=15b$ for b.

Distribute $3a$ onto the inner terms.

$$3ab-3ac=15b$$

Bring both the b terms onto the right side.

$$3ab-3ac=15b$$
$$\underset{-3ab}{}\underset{-3ab}{}$$

$$-3ac=15b-3ab$$

Factor out b.

$$-3ac=b(15-3a)$$

$$\frac{-3ac}{15-3a}=\frac{b(15-3a)}{15-3a}$$

$$\frac{-3ac}{15-3a}=b$$

It is possible to simplify further.

$$\frac{-3ac}{3(5-a)}=b$$

$$\frac{-ac}{5-a}=b$$

3.128 Solve $\dfrac{D}{k}+16=1-c$ for k.

First, eliminate 16 from the left.

$$\frac{D}{k}+16=1-c$$
$$\phantom{\frac{D}{k}+}\underset{-16}{}\underset{-16}{}$$

$$\frac{D}{k}=-15-c$$

$$\frac{D}{k}=\frac{-15-c}{1}$$

Take reciprocals of both sides.

$$\frac{k}{D}=\frac{1}{-15-c}$$

Multiply both sides by D.

$$\frac{k}{D}\times D=\frac{1}{-15-c}\times D$$

$$\frac{kD}{D}=\frac{D}{-15-c}$$

$$k=\frac{D}{-15-c}$$

3.129 $x=-y+5$ and $6x+8y=34$

In the first equation, x is expressed in the form of other terms. Substitute those terms for x in the second equation.

$$6x+8y=34$$
$$6(-y+5)+8y=34$$
$$-6y+30+8y=34$$
$$2y+30=34$$
$$\quad-30\quad-30$$
$$2y=4$$
$$\frac{2y}{2}=\frac{4}{2}$$
$$y=2$$

Now that y is known, plug it in the first equation to find the value of x.

$$x=-y+5$$
$$x=-2+5$$
$$x=3$$

3.130 Assume h as Hillary's height, and d as Donald's height. The first sentence of the problem means $h=d+3$. The second sentence means $h+d=51$. In the first equation, h is expressed as other terms. Substitute those terms in the place of h in the second equation.

$$h+d=51$$
$$(d+3)+d=51$$
$$2d+3=51$$
$$\quad-3\quad-3$$
$$2d=48$$
$$\frac{2d}{2}=\frac{48}{2}$$
$$d=24$$

Now that d is known, plug it in the first equation to find the value of h.

$$h=d+3$$
$$h=24+3$$
$$h=27$$

Hillary is 27 inches tall, and Donald is 24 inches tall. (These seem to be little kids. Any semblance to the names of any persons, living or dead, is *purely coincidental*.)

3.131 $2x+4y=6$ and $3x-y=16$

Eliminate the y terms, their LCM being $4y$. Multiply the second equation by 4.

$$4(3x-y)=4(16)$$
$$12x-4y=64$$
$$+\ 2x+4y=\ 6$$
$$\overline{14x\qquad=70}$$

$$\frac{14x}{14}=\frac{70}{14}$$
$$x=5$$

Now that x is known, plug it in the second equation to find the value of y.

$$3x-y=16$$
$$3(5)-y=16$$
$$15-y=16$$
$$-15\qquad-15$$
$$-y=1$$

Multiply both sides by –1 to reverse the signs.

$$y=-1$$

3.132 Assume an egg weighs e ounces, and a tangerine weighs t ounces. Then the first sentence of the problem means $2e+3t=13$. The second sentence means $3e+8t=30$. Eliminate the e terms, their LCM being $6e$. Multiply the first equation by 3, and the second by –2.

$$3(2e+3t)=3(13)\quad\Big|\quad-2(3e+8t)=-2(30)$$
$$6e+9t=39\qquad\quad -6e-16t=-60$$

Now add the two equations.

$$-6e-16t=-60$$
$$+\ 6e+9t=\ 39$$
$$\overline{\qquad-7t=-21}$$

$$\frac{-7t}{-7}=\frac{-21}{-7}$$
$$t=3$$

Now that t is known, plug it in the first equation to find the value of e.

$$2e+3t=13$$
$$2e+3(3)=13$$
$$2e+9=13$$
$${}_{-9}{}_{-9}$$
$$2e=4$$
$$\frac{2e}{2}=\frac{4}{2}$$
$$e=2$$

3.133 Solve: $|x-8|+3=7$

First, get the modulus by itself.
$$|x-8|+3=7$$
$${}_{-3}{}_{-3}$$
$$|x-8|=4$$

Modulus is positive. Equation is valid. Now solve.

$x-8=4$	$-(x-8)=4$
$x-8=4$	$-x+8=4$
${}_{+8}\quad{}_{+8}$	
$x=12$	$-x+8=4$
	${}_{-8}\quad{}_{-8}$
	$-x=-4$
	$x=4$

3.134 Solve: $14+|2-5x|=6$

First, get the modulus by itself.
$$14+|2-5x|=6$$
$${}_{-14}{}_{-14}$$
$$|2-5x|=-8$$

Modulus is negative. Equation has no solution.

3.135 $\quad 3x^2-2x+4=14-x$
$${}_{+x}{}_{+x}$$
$$3x^2-x+4=14$$
$${}_{-14}\quad{}_{-14}$$
$$3x^2-x-10=0$$
$$3x^2-6x+5x-10=0$$
$$3x(x-2)+5(x-2)=0$$
$$(x-2)(3x+5)=0$$

This is a zero product. Equate each factor to zero.

$x-2=0$	$3x+5=0$
${}_{+2}\quad{}_{+2}$	${}_{-5}\quad{}_{-5}$
$x=2$	$3x=-5$
	$\frac{3x}{3}=\frac{-5}{3}$
	$x=-\frac{5}{3}=-1\frac{2}{3}$

3.136 Comparing $2x^2-5x-6=0$ with $ax^2+bx+c=0$, note that $a=2$, $b=-5$, $c=-6$.

$$x=\frac{-b\pm\sqrt{b^2-4ac}}{2a}$$
$$x=\frac{-(-5)\pm\sqrt{(-5)^2-4(2)(-6)}}{2(2)}$$
$$x=\frac{5\pm\sqrt{25+48}}{4}$$
$$x=\frac{5\pm\sqrt{73}}{4}$$

The answers are $x=\dfrac{5+\sqrt{73}}{4}$ and $x=\dfrac{5-\sqrt{73}}{4}$

Rounded to two decimal digits, the answers are:
$$x=3.39 \text{ and } x=-0.89$$

3.137 Solve: $16^{x+5}=2^{6x-10}$

The bases are not the same. To equate them, write 16 as 2^4.

$$(2^4)^{x+5}=2^{6x-10}$$
$$2^{4x+20}=2^{6x-10} \qquad \text{Refer page 43, law 3.}$$

The bases are the same. Equate the exponents.
$$4x+20=6x-10$$
$${}_{-6x}{}_{-6x}$$
$$-2x+20=-10$$
$${}_{-20}\quad{}_{-20}$$
$$-2x=-30$$
$$\frac{-2x}{-2}=\frac{-30}{-2}$$
$$x=15$$

3.138 Solve: $\sqrt{2x-7}-13=2$

First, isolate the radical.
$$\sqrt{2x-7}-13=2$$
$$\phantom{\sqrt{2x-7}}{}_{+13}\quad{}_{+13}$$
$$\sqrt{2x-7}=15$$

Square both sides to eliminate the radical sign.
$$(\sqrt{2x-7})^2=15^2$$
$$2x-7=15\cdot15$$
$$2x-7=225$$
$${}_{+7}\quad{}_{+7}$$
$$2x=232$$
$$\frac{2x}{2}=\frac{232}{2}$$
$$x=116$$

CHAPTER 4: FUNCTIONS AND GRAPHS

4.001 When $x=4$,

$f(4)=3\cdot(4)-8=12-8=4$

When $x=-2$,

$f(-2)=3\cdot(-2)-8=-6-8=-14$

When $x=0$,

$f(0)=3\cdot(0)-8=0-8=-8$

4.002 $g(6)=6^2+7=36+7=43$

$g(-10)=(-10)^2+7=100+7=107$

$g(1)=1^2+7=1+7=8$

4.003 $F(16)=\dfrac{16+4}{16-6}=\dfrac{20}{10}=2$

$F(22)=\dfrac{22+4}{22-6}=\dfrac{26}{16}=\dfrac{13}{8}=1\dfrac{5}{8}$

4.010 $3x+5y=30$

Solve for y to get the form $y=ax+b$.

$3x+5y=30$
$_{-3x}\qquad _{-3x}$

$5y=30-3x$

$\dfrac{5y}{5}=\dfrac{30-3x}{5}$

$y=\dfrac{30}{5}-\dfrac{3x}{5}$

$y=6-\dfrac{3}{5}x$

$y=-\dfrac{3}{5}x+6$

The *run* is 5. So go 5 steps to the left and right of 0 on the x-axis. That's x-coordinates -5 and $+5$. Calculate the y-coordinates at $x=-5,+5,$ 0 using the equation you have. Then plot those points and draw a line through them. The calculations and graph follow:

x	$y=-\dfrac{3}{5}x+6$
-5	$y=-\dfrac{3}{5}(-5)+6=\dfrac{15}{5}+6=3+6=9$
0	$y=-\dfrac{3}{5}(0)+6=0+6=6$
5	$y=-\dfrac{3}{5}(5)+6=-\dfrac{15}{5}+6=-3+6=3$

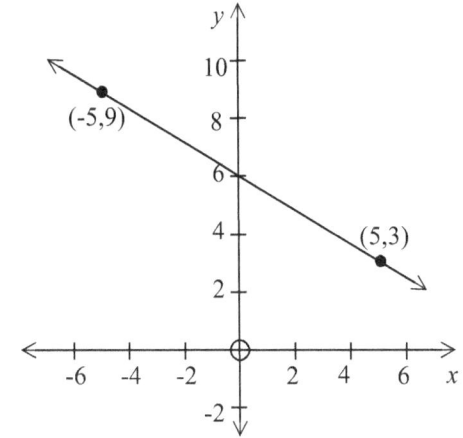

4.011 $y=\dfrac{2}{7}x-5$

The *run* is 7. So go 7 paces to the left and right of 0 on the x-axis. That's x-coordinates -7 and $+7$. Calculate the y-coordinates at $x=-7,+7,$ 0 using the equation you have. Then plot those points and draw a line through them. The calculations and graph follow:

x	$y=\dfrac{2}{7}x-5$
-7	$y=\dfrac{2}{7}(-7)-5=\dfrac{-14}{7}-5=-2-5=-7$
0	$y=\dfrac{2}{7}(0)-5=0-5=-5$
7	$y=\dfrac{2}{7}(7)-5=\dfrac{14}{7}-5=2-5=-3$

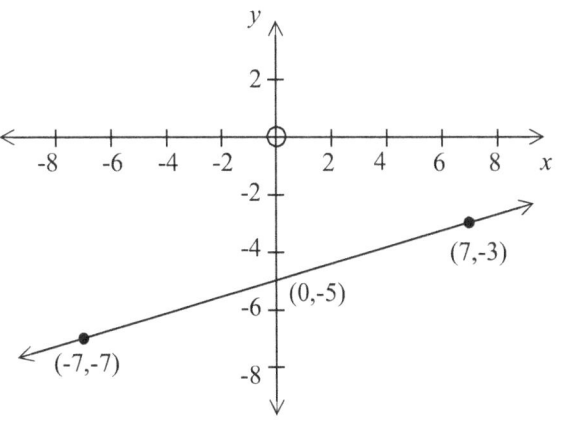

4.012 $f(x)=\dfrac{x}{4}+2$ means $y=\dfrac{1}{4}x+2$.

The *run* is 4. So go 4 steps to the left and right of 0 on the *x*-axis. That's *x*-coordinates −4 and +4. Calculate the *y*-coordinates at $x=-4,+4,\ 0$ using the equation you have. Then mark those points and draw a line through them. The calculations and graph follow:

x	$y=\dfrac{1}{4}x+2$
−4	$y=\dfrac{1}{4}(-4)+2=\dfrac{-4}{4}+2=-1+2=1$
0	$y=\dfrac{1}{4}(0)+2=0+2=2$
4	$y=\dfrac{1}{4}(4)+2=\dfrac{4}{4}+2=1+2=3$

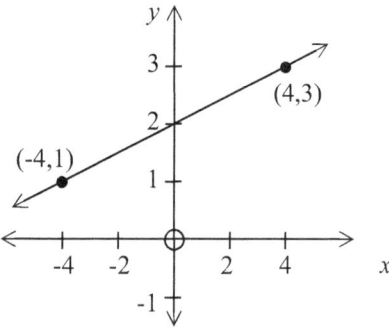

4.013 $g(x)=-\dfrac{3}{4}x-3$ means $y=-\dfrac{3}{4}x-3$.

The *run* is 4. So go 4 paces to the left and right of 0 on the *x*-axis. That's *x*-coordinates −4 and +4. Calculate the *y*-coordinates at $x=-4,+4,\ 0$ using the equation you have. Then plot those points and draw a line through them. The calculations and graph follow:

x	$y=-\dfrac{3}{4}x-3$
−4	$y=-\dfrac{3}{4}(-4)-3=\dfrac{12}{4}-3=3-3=0$
0	$y=-\dfrac{3}{4}(0)-3=0-3=-3$
4	$y=-\dfrac{3}{4}(4)-3=-\dfrac{12}{4}-3=-3-3=-6$

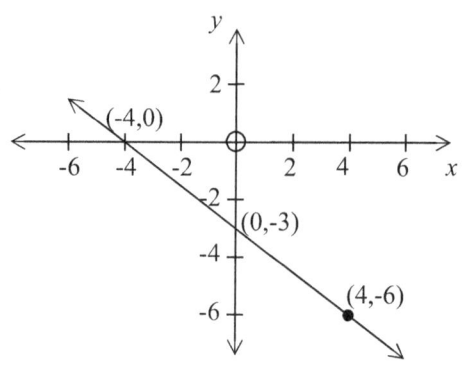

4.014 $x=\dfrac{1}{5}y+2$

Solve for *y* to get the form $y=ax+b$.

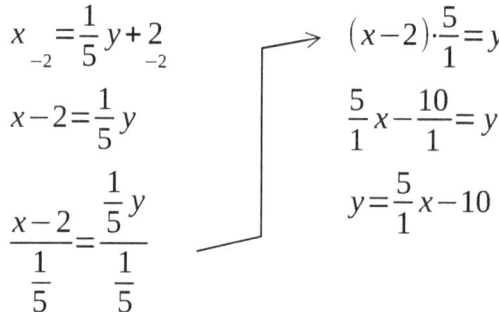

$$x_{-2}=\frac{1}{5}y+2_{-2}$$

$$x-2=\frac{1}{5}y$$

$$\frac{x-2}{\frac{1}{5}}=\frac{\frac{1}{5}y}{\frac{1}{5}}$$

$$(x-2)\cdot\frac{5}{1}=y$$

$$\frac{5}{1}x-\frac{10}{1}=y$$

$$y=\frac{5}{1}x-10$$

The *run* is 1. So go 1 step to the left and right of 0 on the *x*-axis. That's *x*-coordinates −1 and +1. Calculate the *y*-coordinates at $x=-1,+1,\ 0$ using a simpler form $y=5x-10$. Then plot those points and draw a line through them. The calculations and graph follow:

x	$y=5x-10$
−1	$y=5(-1)-10=-5-10=-15$
0	$y=5(0)-10=0-10=-10$
1	$y=5(1)-10=5-10=-5$

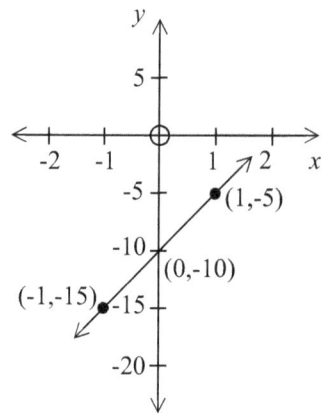

4.015 $q(x)=-5x$ means $y=-5x$ which means $y=-\dfrac{5}{1}x+0$. The *run* is 1. So go 1 pace to the left and right of 0 on the x-axis. That would get you the x-coordinates −1 and +1. Calculate the y-coordinates $x=-1,+1,$ 0 using the simpler form $y=-5x$. Then plot those points and draw a line through them. The calculations and graph follow:

x	$y=-5x$
−1	$y=-5(-1)=5$
0	$y=-5(0)=0$
1	$y=-5(1)=-5$

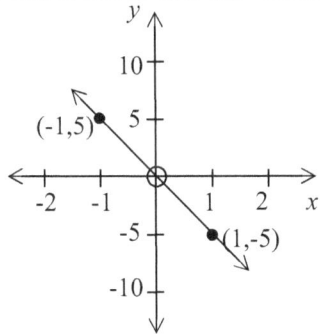

4.021 First, calculate the y-coordinates at $x=-1,$ +1, and 0. Then plot the points and draw a smooth curve through them. Table and graph follow. The domain is $(-\infty,\infty)$, and the range is $(0,\infty)$.

x	$y=6\cdot\left(\dfrac{1}{3}\right)^x$
−1	$y=6\cdot\left(\dfrac{1}{3}\right)^{-1}=6\cdot\left(\dfrac{1^{-1}}{3^{-1}}\right)=6\cdot\dfrac{3}{1}=\dfrac{18}{1}=18$
0	$y=6\cdot\left(\dfrac{1}{3}\right)^{0}=6\cdot\dfrac{1^0}{3^0}=6\cdot\dfrac{1}{1}=6$
1	$y=6\cdot\left(\dfrac{1}{3}\right)^{1}=6\cdot\dfrac{1^1}{3^1}=6\cdot\dfrac{1}{3}=\dfrac{6}{3}=2$

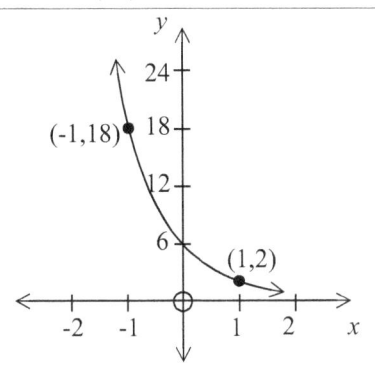

4.022 $f(x)=-15\cdot(3)^{x-1}$

Solve the function to get the form $y=a\cdot b^x$.

$$y=-15\cdot(3)^{x-1}$$

$$y=-15\cdot\dfrac{3^x}{3^1} \qquad \text{Refer page 43, law 2.}$$

$$y=\dfrac{-15\cdot3^x}{3}$$

$$y=-5\cdot(3)^x$$

Now calculate the y-coordinates at $x=-1,$ +1, and 0. Then plot the points and draw a smooth curve through them. Table and graph follow. The domain is $(-\infty,\infty)$, and the range is $(-\infty,0)$. Smaller value first.

x	$y=-5\cdot(3)^x$
−1	$y=-5\cdot(3)^{-1}=-5\cdot\dfrac{1}{3}=-\dfrac{5}{3}=-1.\overline{6}$
0	$y=-5\cdot(3)^{0}=-5\cdot1=-5$
1	$y=-5\cdot(3)^{1}=-5\cdot3=-15$

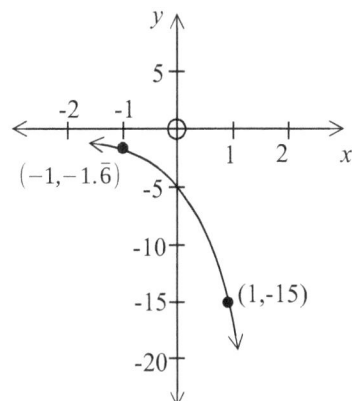

4.023 First, calculate the y-coordinates at $x=-1,$ +1, and 0. Then plot the points and draw a smooth curve through them. Table is shown below and graph is on the next page. The domain is $(-\infty,\infty)$, and the range is $(-\infty,0)$.

x	$y=-(4)^x$
−1	$y=-(4)^{-1}=-\left(\dfrac{1}{4^1}\right)=-\dfrac{1}{4}=-0.25$
0	$y=-(4)^{0}=-1$
1	$y=-(4)^{1}=-4$

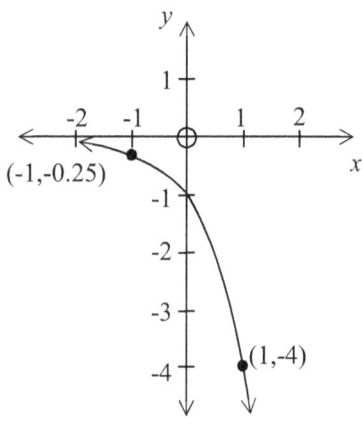

(-1,-0.25)

(1,-4)

4.024 $q(x)=-8\cdot(4)^{-x}$

Solve the function to get the form $y=a\cdot b^x$.

$$y=-8\cdot(4)^{-x}$$

$$y=-8\cdot\frac{1}{4^x}$$ Refer page 43, law 4.

$$y=-8\cdot\left(\frac{1^x}{4^x}\right)$$ because $1^x=1$

$$y=-8\cdot\left(\frac{1}{4}\right)^x$$ Refer page 43, law 8.

$$y=-8\cdot(0.25)^x$$

Now calculate the y-coordinates at $x=-1$, $+1$, and 0. Then plot the points and draw a smooth curve through them. Table and graph follow. The domain is $(-\infty,\infty)$, and the range is $(-\infty,0)$.

x	$y=-8\cdot(0.25)^x$
-1	$y=-8\cdot(0.25)^{-1}=-8\cdot\dfrac{1}{0.25}=-8\cdot4=-32$
0	$y=-8\cdot(0.25)^0=-8\cdot1=-8$
1	$y=-8\cdot(0.25)^1=-8\cdot(0.25)=-2$

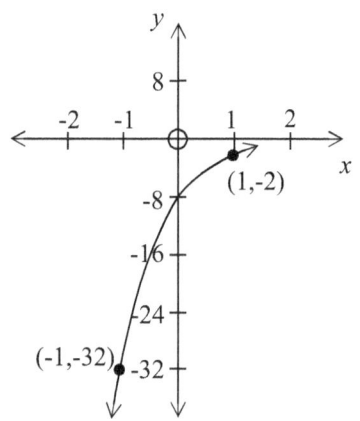

(1,-2)

(-1,-32)

4.025 $y=4\cdot(5)^{x+2}$

Solve the function to get the form $y=a\cdot b^x$.

$$y=4\cdot(5)^{x+2}$$

$$y=4\cdot5^x\cdot5^2$$ Refer page 43, law 1.

$$y=4\cdot5^x\cdot25$$

$$y=4\cdot25\cdot5^x$$

$$y=100\cdot(5)^x$$

Now calculate the y-coordinates at $x=-1$, $+1$, and 0. Then plot the points and draw a smooth curve through them. Table and graph follow. The domain is $(-\infty,\infty)$, and the range is $(0,\infty)$.

x	$y=100\cdot(5)^x$
-1	$y=100\cdot(5)^{-1}=100\cdot\dfrac{1}{5}=\dfrac{100}{5}=20$
0	$y=100\cdot(5)^0=100\cdot1=100$
1	$y=100\cdot(5)^1=100\cdot5=500$

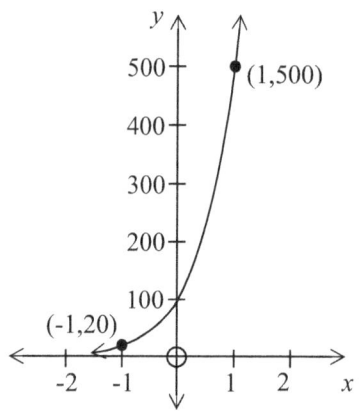

(1,500)

(-1,20)

4.032 $f(x)=-3x^2+6x-1$

Comparing it with the form $y=ax^2+bx+c$, note that $a=-3$, $b=6$, $c=-1$.

The x-coordinate of the vertex is

$$\frac{-b}{2a}=\frac{-6}{2\cdot(-3)}=\frac{-6}{-6}=1$$

To find the y-coordinate of the vertex, plug 1 for x into the quadratic function:

$$y=-3(1)^2+6(1)-1=-3+6-1=2$$

Mark the vertex at (1,2) and the y-intercept at $(0,c)$ which is (0,–1). The mirror image of the y-intercept will be horizontally twice as far as the vertex, and so its x-coordinate will be 2. But its y-coordinate will be the same as that of the

y-intercept, –1. Mark the mirror point at (2,–1). Figure below shows the parabola with its domain $(-\infty,\infty)$ and range $(-\infty,2]$.

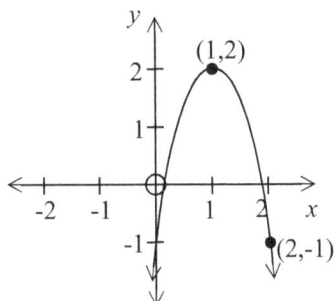

4.033 $y=(x-1)(x+7)$

The function is not in the standard form $y=ax^2+bx+c$. So first simplify it.

$y=x^2+7x-1x-7$

$y=x^2+6x-7$

Comparing it with the standard form, note that $a=1$, $b=6$, $c=-7$.

The x-coordinate of the vertex is

$$\frac{-b}{2a}=\frac{-6}{2\cdot(1)}=\frac{-6}{2}=-3$$

To find the y-coordinate of the vertex, plug –3 for x into the quadratic function:

$$y=(-3)^2+6(-3)-7=9-18-7=-16$$

Mark the vertex at (–3,–16) and the y-intercept at $(0,c)$ which is (0,–7). The mirror image of the y-intercept will be horizontally twice as far as the vertex, and so its x-coordinate will be –6. But its y-coordinate will be the same as that of the y-intercept, –7. Mark the mirror point at (–6,–7).

Below is drawn the parabola with its domain $(-\infty,\infty)$ and range $[-16,\infty)$.

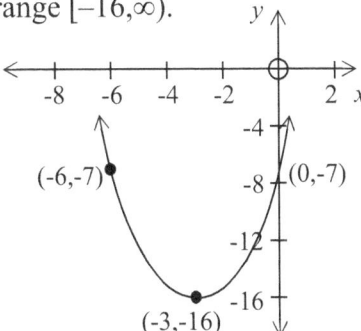

4.034 $j(x)=4x^2$ Comparing the function with $y=ax^2+bx+c$, note that $a=4$, $b=0$, $c=0$.

The x-coordinate of the vertex is

$$\frac{-b}{2a}=\frac{-(0)}{2\cdot(4)}=\frac{0}{8}=0$$

To find the y-coordinate of the vertex, plug 0 for x into the function: $y=4(0)^2=4(0)=0$

The vertex is at (0,0) and so is the y-intercept. In such cases, find the y-coordinates of two neighboring points, say at $x=-1$ and $x=+1$.

At $x=-1$, $y=4(-1)^2=4(1)=4$.

The point is (–1,4).

At $x=1$, $y=4(1)^2=4(1)=4$.

The point is (1,4).

Figure below shows the parabola with its domain $(-\infty,\infty)$ and range $(0,\infty)$.

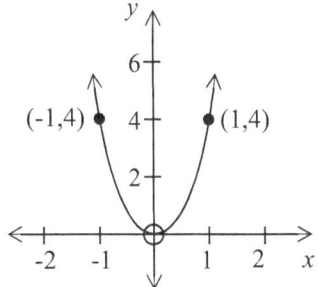

4.035 $y=(x+4)^2-8$

The function is not in the standard form $y=ax^2+bx+c$. So first simplify it.

$y=(x+4)^2-8$

$y=(x+4)(x+4)-8$

$y=x^2+4x+4x+16-8$

$y=x^2+8x+8$

Comparing it with the standard form, note that $a=1$, $b=8$, $c=8$.

The x-coordinate of the vertex is

$$\frac{-b}{2a}=\frac{-8}{2\cdot(1)}=\frac{-8}{2}=-4$$

To find the y-coordinate of the vertex, plug –4 for x into the quadratic function:

$$y=(-4)^2+8(-4)+8=16-32+8=-8$$

Mark the vertex at (–4,–8) and the y-intercept at $(0,c)$ which is (0,8). The mirror image of the y-intercept will be horizontally twice as far as the vertex, and so its x-coordinate will be –8. But its y-coordinate will be the same as that of the y-intercept, 8. Mark the mirror point at (–8,8).

Figure at the top left corner of the next page shows the parabola with its domain $(-\infty,\infty)$ and range $[-8,\infty)$.

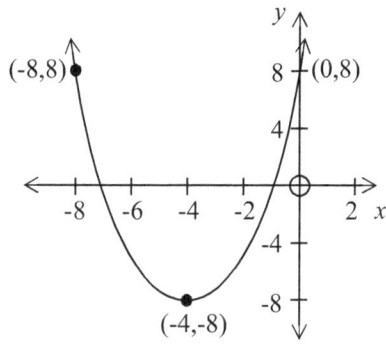

(-8,8) 8 (0,8)
-8 -6 -4 -2 2 x
-4
-8
(-4,-8)

4.036 $k(x)=25-x^2$ The function may be written as $y=-x^2+0x+25$. Comparing it with $y=ax^2+bx+c$, note that $a=-1, b=0, c=25$.
The x-coordinate of the vertex is
$$\frac{-b}{2a}=\frac{-(0)}{2\cdot(-1)}=\frac{0}{-2}=0$$
To find the y-coordinate of the vertex, plug 0 for x into the function: $y=-(0)^2+0(0)+25=25$
The vertex is at $(0,25)$ as is the y-intercept. In such cases, find the y-coordinates of two neighboring points, say at $x=-1$ and $x=+1$.
At $x=-1$, $y=-(-1)^2+0(-1)+25=-1+25=24$
The point is $(-1,24)$.
At $x=1$, $y=-(1)^2+0(1)+25=-1+25=24$
The point is $(1,24)$.
Figure below shows the parabola with its domain $(-\infty,\infty)$ and range $(-\infty,25]$.

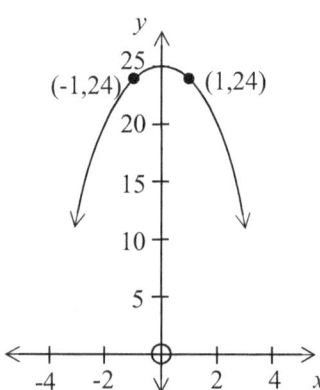

25
(-1,24) (1,24)
20
15
10
5
-4 -2 2 4 x

4.037 $y=x-\frac{1}{2}x^2$ The function may be written as $y=-0.5x^2+x+0$. Comparing it with $y=ax^2+bx+c$, note that $a=-0.5, b=1, c=0$.
The x-coordinate of the vertex is
$$\frac{-b}{2a}=\frac{-1}{2\cdot(-0.5)}=\frac{-1}{-1}=1$$

To find the y-coordinate of the vertex, plug 1 for x into the function:
$$y=-0.5(1)^2+1+0=-0.5+1=0.5$$
Mark the vertex at $(1, 0.5)$ and the y-intercept at $(0,c)$ which is $(0,0)$. The mirror image of the y-intercept will be horizontally twice as far as the vertex, and so its x-coordinate will be 2. But its y-coordinate will be the same as that of the y-intercept, 0. Mark the mirror point at $(2,0)$.
Figure below shows the parabola with its domain $(-\infty,\infty)$ and range $(-\infty,0.5]$.

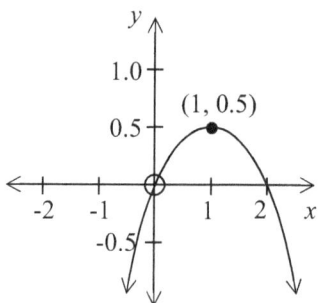

y
1.0
(1, 0.5)
0.5
-2 -1 1 2 x
-0.5

4.042 $y=-7|x|$ The function may be written as $y=-7|x-0|+0$. Comparing it with the standard form $y=a|x-b|+c$ note that $a=-7$, $b=0$, $c=0$. The vertex is at (b,c) which is $(0,0)$.
Find the y-intercept, whose x-coordinate is always 0, and whose y-coordinate can be calculated by plugging 0 for x into the function:
$$y=-7|0|=-7\cdot0=0$$
So the y-intercept is at $(0,0)$, same as vertex. In such cases, find the y-coordinates of two neighboring points, say at $x=-1$ and $x=+1$.
At $x=-1$, $y=-7|-1|=-7\cdot1=-7$
The point is $(-1,-7)$.
At $x=1$, $y=-7|1|=-7\cdot1=-7$
The point is $(1,-7)$.
Draw rays from the vertex through these two points, as shown below. The domain is $(-\infty,\infty)$ and the range is $(-\infty,0]$.

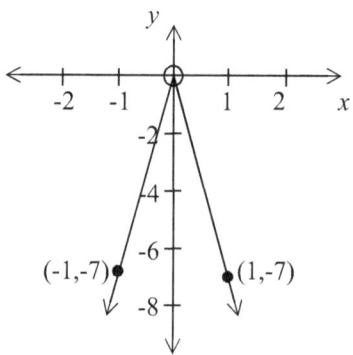

y
-2 -1 1 2 x
-2
-4
-6
(-1,-7) (1,-7)
-8

4.043 $h(x)=|x-2|$ The function may be written as $y=1\cdot|x-2|+0$. Comparing it with the standard form $y=a|x-b|+c$ note that $a=1$, $b=2$, $c=0$. The vertex is at (b,c) which is $(2,0)$.

Find the y-intercept, whose x-coordinate is always 0, and whose y-coordinate can be calculated by plugging 0 for x into the function:

$y=1\cdot|0-2|+0=|-2|=2$

So the y-intercept is at $(0,2)$. Its mirror image would be on the other side of the vertex, with x-coordinate twice that of vertex, or 4, and y-coordinate same as that of y-intercept, or 2. The point is $(4,2)$.

Draw rays from the vertex through these two points, as shown below. The domain is $(-\infty,\infty)$ and range is $[0,\infty)$.

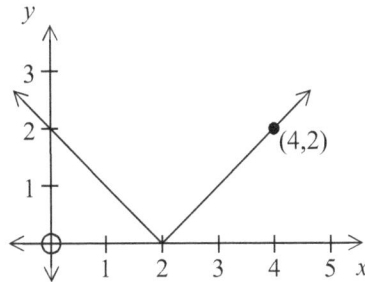

4.044 $f(x)=-3|x|+10$ The function may be written as $y=-3|x-0|+10$. Comparing it with the standard form $y=a|x-b|+c$ note that $a=-3$, $b=0$, $c=10$. The vertex is at (b,c) which is $(0,10)$.

Find the y-intercept, whose x-coordinate is always 0, and whose y-coordinate can be calculated by plugging 0 for x into the function:

$y=-3|0-0|+10=-3\cdot0+10=10$

So the y-intercept is at $(0,10)$, same as vertex. In such cases, find the y-coordinates of two neighboring points, say at $x=-1$ and $x=+1$.

At $x=-1$, $y=-3\cdot|-1-0|+10=-3+10=7$.

The point is $(-1,7)$.

At $x=1$, $y=-3\cdot|1-0|+10=-3+10=7$.

The point is $(1,7)$.

Draw rays from the vertex through these two points, as shown at the top right corner of this page. The domain is $(-\infty,\infty)$ and range is $(-\infty,10]$.

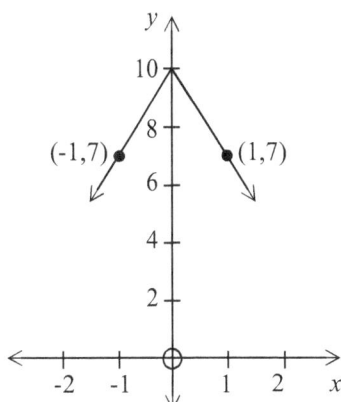

4.045 $y=5|x+2|-4$ Comparing the function with the standard form $y=a|x-b|+c$ note that $a=5$, $b=-2$, $c=-4$. The vertex is at (b,c) which is $(-2,-4)$.

Find the y-intercept, whose x-coordinate is always 0, and whose y-coordinate can be calculated by plugging 0 for x into the function:

$y=5|0+2|-4=5|2|-4=10-4=6$

So the y-intercept is at $(0,6)$. Its mirror image would be on the other side of the vertex, with x-coordinate twice that of vertex, or -4, and y-coordinate same as that of y-intercept, or 6. The point is $(-4,6)$.

Draw rays from the vertex through these two points. The domain is $(-\infty,\infty)$ and range is $[-4,\infty)$.

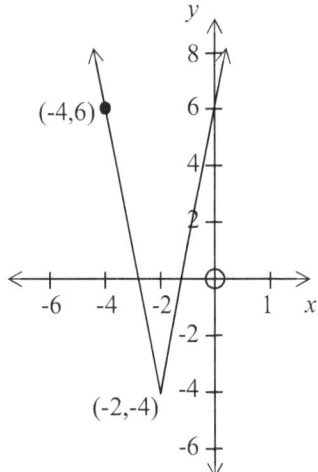

4.049 To find $f(-50)$, put -50 in the place of x in $f(x)$.

$f(x)=5x$

$f(-50)=5(-50)=-250$

To find $g(24)$, write 24 for x in $g(x)$.

$g(x)=x+7$

$g(24)=24+7=31$

To find $f(g(x))$, put $g(x)$ in the place of x in the original function $f(x)$:

$f(x)=5x$
$f(g(x))=5\cdot g(x)$

But since $g(x)$ is given as $x+7$, put that in the above composite function:

$f(g(x))=5\cdot(x+7)$
$f(g(x))=5x+35$ This step is optional.

To find $g(f(a))$, first find $g(f(x))$. Then plug a into it. To first find $g(f(x))$, put $f(x)$ in the place of x in the original function $g(x)$:

$g(x)=x+7$
$g(f(x))=f(x)+7$

But $f(x)$ is given as $5x$. So put that in the above composite function:

$g(f(x))=5x+7$
$g(f(a))=5a+7$

4.050 To find $g(-3)$, put -3 for x in $g(x)$.

$g(x)=x^2-7$
$g(-3)=(-3)^2-7=9-7=2$

To find $h(-10)$, put -10 for x in $h(x)$.

$h(x)=2x$
$h(-10)=2(-10)=-20$

To find $g(h(4))$, first find $g(h(x))$. Then plug 4 into it. To first find $g(h(x))$, put $h(x)$ in the place of x in the original function $g(x)$:

$g(x)=x^2-7$
$g(h(x))=[\ h(x)\ ]^2-7$

But $h(x)$ is given as $2x$. So put that in the above composite function:

$g(h(x))=(2x)^2-7$
$g(h(x))=2^2x^2-7$
$g(h(x))=4x^2-7$
$g(h(4))=4(4)^2-7=4\cdot16-7=64-7=57$

To find $h(g(x))$, put $g(x)$ in the place of x in the original function $h(x)$:

$h(x)=2x$
$h(g(x))=2\cdot g(x)$

But $g(x)$ is given as x^2-7. So put that in the above composite function:

$h(g(x))=2(x^2-7)$
$h(g(x))=2x^2-14$ This step is optional.

4.051 To find $f(h(4))$, first find $f(h(x))$. Then plug 4 into it. To first find $f(h(x))$, put $h(x)$ in the place of x in the original function $f(x)$:

$f(x)=x+2$
$f(h(x))=h(x)+2$

But $h(x)$ is given as 3^x. So put that in the above composite function:

$f(h(x))=3^x+2$
$f(h(4))=3^4+2=3\cdot3\cdot3\cdot3+2=81+2=83$

To find $h(f(c))$, first find $h(f(x))$. Then plug c into it. To first find $h(f(x))$, put $f(x)$ in the place of x in the original function $h(x)$:

$h(x)=3^x$
$h(f(x))=3^{f(x)}$

But $f(x)$ is given as $x+2$. So put that in the above composite function:

$h(f(x))=3^{x+2}$
$h(f(x))=3^x\cdot3^2$ Refer page 43, law 1.
$h(f(x))=3^x\cdot9$
$h(f(x))=9\cdot(3)^x$
$h(f(c))=9\cdot(3)^c$

4.055 First, write the function as $y=\dfrac{1}{4}x^2$.

Then exchange x and y to get $x=\dfrac{1}{4}y^2$.

Next, solve for y:

$$\frac{x}{\frac{1}{4}}=\frac{\frac{1}{4}y^2}{\frac{1}{4}}$$

$$x\cdot\frac{4}{1}=y^2$$

$$4x=y^2$$
$$\sqrt{4x}=\sqrt{y^2}$$
$$2\sqrt{x}=y$$

Replace y with $p^{-1}(x)$.

$2\sqrt{x}=p^{-1}(x)$
$p^{-1}(x)=2\sqrt{x}$

4.056 First, write the function as $y=-6x+7$.
Now exchange x and y to get $x=-6y+7$.
Next, solve for y:

$$x = -6y + 7$$
$$_{-7} _{-7}$$
$$x - 7 = -6y$$
$$\frac{x-7}{-6} = \frac{-6y}{-6}$$
$$\frac{x}{-6} - \frac{7}{-6} = y$$
$$-\frac{x}{6} + \frac{7}{6} = y$$

Replace y with $f^{-1}(x)$.

$$-\frac{x}{6} + \frac{7}{6} = f^{-1}(x)$$

$$f^{-1}(x) = -\frac{x}{6} + \frac{7}{6}$$

4.057 First, write the function as $y = 16x^2$.
Then exchange x and y to get $x = 16y^2$.
Next, solve for y:

$$\frac{x}{16} = \frac{16y^2}{16}$$

$$\frac{x}{16} = y^2$$

$$\sqrt{\frac{x}{16}} = \sqrt{y^2}$$

$$\frac{\sqrt{x}}{\sqrt{16}} = y$$

$$\frac{\sqrt{x}}{4} = y$$

Replace y with $q^{-1}(x)$.

$$\frac{\sqrt{x}}{4} = q^{-1}(x)$$

$$q^{-1}(x) = \frac{\sqrt{x}}{4}$$

4.058 To find $f(-2)$, put -2 for x in $f(x)$:

$$f(x) = 3x^2 - 8$$
$$f(-2) = 3(-2)^2 - 8$$
$$f(-2) = 3(4) - 8 = 12 - 8 = 4$$

To find $f(0)$, put 0 for x in $f(x)$:

$$f(0) = 3(0)^2 - 8 = 0 - 8 = -8$$

To find $f(c)$, put c for x in $f(x)$:

$$f(c) = 3c^2 - 8$$

To find $f(3)$, put 3 for x in $f(x)$:

$$f(3) = 3(3)^2 - 8 = 3(9) - 8 = 27 = 8 = 19$$

4.059 To find $f(10)$, put 10 for x in $f(x)$:

$$f(x) = 2x - 1$$
$$f(10) = 2(10) - 1 = 20 - 1 = 19$$

To find $f(-3)$, put -3 for x in $f(x)$:

$$f(-3) = 2(-3) - 1 = -6 - 1 = -7$$

To find $f(g(x))$, put $g(x)$ in the place of x in the original function $f(x)$:

$$f(x) = 2x - 1$$
$$f(g(x)) = 2 \cdot g(x) - 1$$

But since $g(x)$ is given as $3x^2$, put that in the above composite function:

$$f(g(x)) = 2 \cdot 3x^2 - 1$$
$$f(g(x)) = 6x^2 - 1$$

To find $g(f(0))$, first find $g(f(x))$. Then plug 0 into it. To first find $g(f(x))$, put $f(x)$ in the place of x in the original function $g(x)$:

$$g(x) = 3x^2$$
$$g(f(x)) = 3[f(x)]^2$$

But since $f(x)$ is given as $2x - 1$, put that in the above composite function:

$$g(f(x)) = 3(2x - 1)^2$$
$$g(f(0)) = 3(2 \cdot 0 - 1)^2$$
$$g(f(0)) = 3(0 - 1)^2 = 3(-1)^2 = 3(1) = 3$$

4.060 The function may be written as $y = \frac{2}{1}x + 5$. The run is 1. So go 1 pace to the left and right of 0 on the x-axis. That's x-coordinates -1 and $+1$. Calculate the y-coordinates at $x = -1$, $+1$, and 0. Then plot those points and draw a line through them. Table and graph follow:

x	$y = 2x + 5$
-1	$y = 2 \cdot (-1) + 5 = -2 + 5 = 3$
0	$y = 2 \cdot (0) + 5 = 0 + 5 = 5$
1	$y = 2 \cdot (1) + 5 = 2 + 5 = 7$

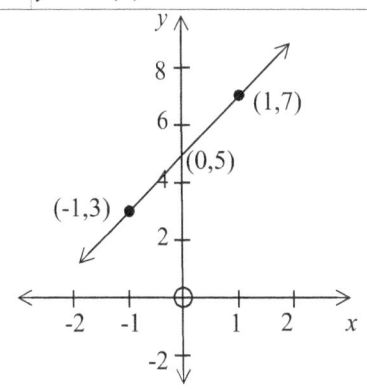

4.061 The function may be written as $y=-\frac{1}{4}x+6$. The run is 4. So go 4 paces to the left and right of 0 on x-axis. That's x-coordinates -4 and $+4$. Calculate the y-coordinates at $x=-4$, $+4$ and 0. Then plot those points and draw a line through them. Table and graph follow:

x	$y=-\frac{1}{4}x+6$
-4	$y=-\frac{1}{4}\cdot(-4)+6=1+6=7$
0	$y=-\frac{1}{4}\cdot0+6=0+6=6$
4	$y=-\frac{1}{4}\cdot4+6=-1+6=5$

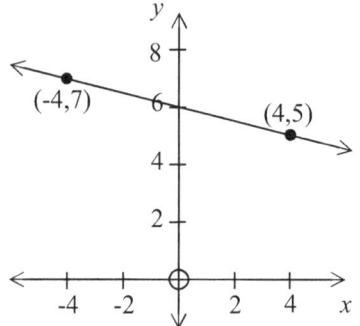

4.062 The function is not in the standard form $y=ax^2+bx+c$. So first simplify it.
$$y=(x+5)(3-x)$$
$$y=3x-x^2+15-5x$$
$$y=-x^2-2x+15$$

Comparing with the standard form, note that $a=-1$, $b=-2$, $c=15$.

The x-coordinate of the vertex is
$$\frac{-b}{2a}=\frac{-(-2)}{2\cdot(-1)}=\frac{2}{-2}=-1$$

To find the y-coordinate of the vertex, plug -1 for x into the quadratic function.
$$y=-(-1)^2-2(-1)+15$$
$$y=-1+2+15=16$$

Mark the vertex at $(-1,16)$ and the y-intercept at $(0,c)$ which is $(0,15)$. The mirror image of the y-intercept will be horizontally twice as far as the vertex, and so its x-coordinate will be -2. But its y-coordinate will be the same as that of the y-intercept, 15. Mark the mirror point at $(-2,15)$. Figure at the top right of this page shows the parabola with domain $(-\infty,\infty)$ and range $(-\infty,16]$.

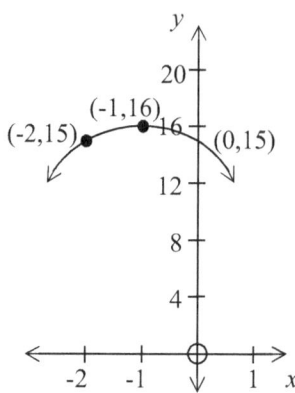

4.063 The function is not in the standard form $y=ax^2+bx+c$. So first simplify it.
$$y=(x-1)^2+11$$
$$y=(x-1)(x-1)+11$$
$$y=x^2-1x-1x+1+11$$
$$y=x^2-2x+12$$

Comparing with the standard form, note that $a=1$, $b=-2$, $c=12$.

The x-coordinate of the vertex is
$$\frac{-b}{2a}=\frac{-(-2)}{2\cdot(1)}=\frac{2}{2}=1$$

To find the y-coordinate of the vertex, plug 1 for x into the quadratic function.
$$y=1^2-2(1)+12$$
$$y=1-2+12=11$$

Mark the vertex at $(1,11)$ and the y-intercept at $(0,c)$ which is $(0,12)$. The mirror image of the y-intercept will be horizontally twice as far as the vertex, and so its x-coordinate will be 2. But its y-coordinate will be the same as that of the y-intercept, 12. Mark the mirror point at $(2,12)$. Shown below is the parabola with domain $(-\infty,\infty)$ and range $[11,\infty)$.

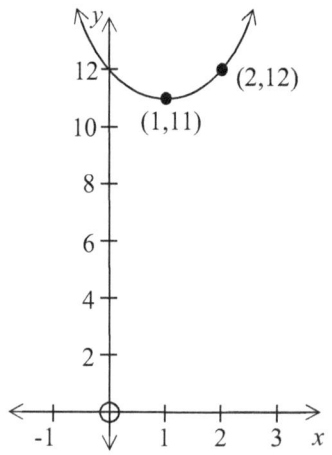

4.064 First calculate the y-coordinates at $x=-1$, $+1$, and 0 using the function which may be written as $y=2\cdot(0.5)^x$ Then plot the points and draw a smooth curve through them. Table and graph follow. The domain of the function is $\mathbb{R}$, or $(-\infty,\infty)$, and the range is $(0,\infty)$.

x	$y=2\cdot(0.5)^x$
-1	$y=2\cdot(0.5)^{-1}=2\cdot\dfrac{1}{0.5}=2\cdot2=4$
0	$y=2\cdot(0.5)^{0}=2\cdot1=2$
1	$y=2\cdot(0.5)^{1}=2\cdot(0.5)=1$

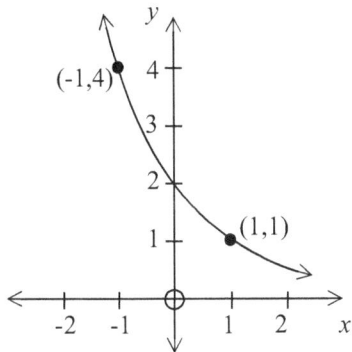

4.065 First calculate the y-coordinates at $x=-1$, $+1$, and 0 using the function which may be written as $y=-5\cdot(2)^x$ Then plot the points and draw a smooth curve through them. Table and graph are shown below. The domain of the function is $(-\infty,\infty)$, and the range is $(-\infty,0)$.

x	$y=-5\cdot(2)^x$
-1	$y=-5\cdot(2)^{-1}=-5\cdot\dfrac{1}{2}=-\dfrac{5}{2}=-2.5$
0	$y=-5\cdot(2)^{0}=-5\cdot1=-5$
1	$y=-5\cdot2^{1}=-5\cdot2=-10$

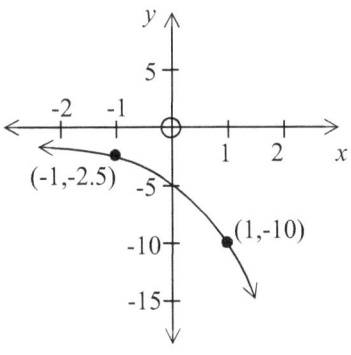

4.066 The function may be written as $y=4\cdot|x+7|+0$. Comparing it with the standard form $y=a|x-b|+c$ note that $a=4$, $b=-7$, $c=0$. The vertex is at (b,c) which is $(-7,0)$.
Now find the y-intercept, whose x-coordinate is always 0, and whose y-coordinate can be calculated by plugging 0 for x into the function:
$y=4\cdot|0+7|+0=4|7|=28$
So the y-intercept is at $(0,28)$. Its mirror image would be on the other side of the vertex, with x-coordinate twice that of vertex, or -14, and y-coordinate same as that of y-intercept, or 28. The point is $(-14,28)$.
Draw rays from the vertex through these two points. Domain is $(-\infty,\infty)$ and range is $[0,\infty)$.

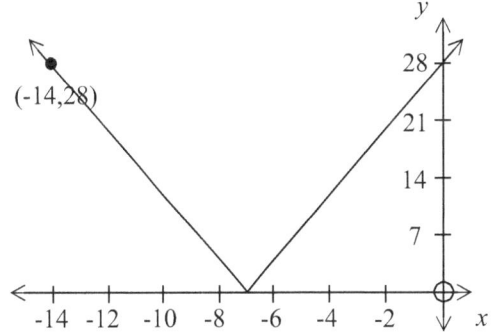

4.067 The function may be written as $y=-5\cdot|x-0|+8$. Comparing it with the standard form $y=a|x-b|+c$ note that $a=-5$, $b=0$, $c=8$. The vertex is at (b,c) which is $(0,8)$.
Now find the y-intercept, whose x-coordinate is always 0, and whose y-coordinate can be calculated by plugging 0 for x into the function:
$y=-5\cdot|0|+8=0+8=8$
So the y-intercept is at $(0,8)$, same as vertex. In such cases, find the y-coordinates of two neighboring points, say at $x=-1$ and $x=+1$.
At $x=-1$, $y=-5|-1|+8=-5+8=3$
The point is $(-1,3)$.
At $x=1$, $y=-5|1|+8=-5+8=3$.
The point is $(1,3)$.
Draw rays from the vertex through these two points, as shown in the figure at the top left corner of the next page. Domain is $(-\infty,\infty)$ and range is $(-\infty,8]$. Smaller value first.

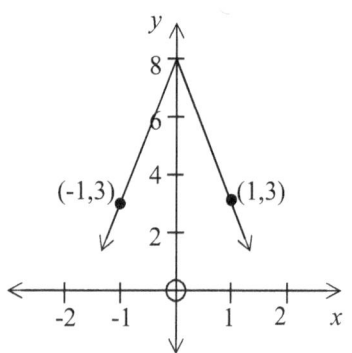

4.068 First, write the function as $y=25x^2$.
Then exchange x and y to get $x=25y^2$.
Next, solve for y:

$$\frac{x}{25}=\frac{25y^2}{25}$$

$$\frac{x}{25}=y^2$$

$$\sqrt{\frac{x}{25}}=\sqrt{y^2}$$

$$\frac{\sqrt{x}}{\sqrt{25}}=y$$

$$\frac{\sqrt{x}}{5}=y$$

Replace y with $h^{-1}(x)$.

$$\frac{\sqrt{x}}{5}=h^{-1}(x)$$

$$h^{-1}(x)=\frac{\sqrt{x}}{5}$$

4.069 First, write the function as $y=\frac{1}{3}x+1$.

Then exchange x and y to get $x=\frac{1}{3}y+1$.

Next, solve for y:

$$x\underset{-1}{=}\frac{1}{3}y\underset{-1}{+1}$$

$$x-1=\frac{1}{3}y$$

$$\frac{x-1}{\frac{1}{3}}=\frac{\frac{1}{3}y}{\frac{1}{3}}$$

$$(x-1)\cdot\frac{3}{1}=y$$

$$3x-3=y$$

Replace y with $f^{-1}(x)$.

$$3x-3=f^{-1}(x)$$

$$f^{-1}(x)=3x-3$$

CHAPTER 5: INEQUALITIES

5.008 $\quad 4x+13\geq 6$
$$\underset{-13}{}\underset{-13}{}$$

$$4x\geq -7$$

$$x\geq -\frac{7}{4} \quad \text{which is} \quad \left[-\frac{7}{4},\infty\right).$$

5.009 $\quad -x+7\geq 5x+1$
$$\underset{-5x}{}\quad\underset{-5x}{}$$

$$-6x+7\geq 1$$

$$-6x+7\geq 1$$
$$\underset{-7}{}\underset{-7}{}$$

$$-6x\geq -6$$

$$\frac{-6x}{-6}\leq\frac{-6}{-6}$$

Division by a negative number reversed the inequality sign.

$$x\leq 1 \quad \text{which is the interval } (-\infty,1].$$

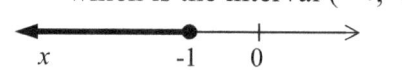

5.010 $\quad -2x+10\leq 8-4x$
$$\underset{+4x}{}\quad\underset{+4x}{}$$

$$2x+10\leq 8$$

$$2x+10\leq 8$$
$$\underset{-10}{}\underset{-10}{}$$

$$2x\leq -2$$

$$\frac{2x}{2}\leq\frac{-2}{2}$$

$$x\leq -1 \quad \text{which is the interval } (-\infty,-1].$$

5.011 $\quad x-3(x-3)>2x+5$

$$x-3x+9>2x+5$$

$$-2x+9>2x+5$$

$$-2x+9>2x+5$$
$$\underset{-2x}{}\quad\underset{-2x}{}$$

$$-4x+9>5$$

$$-4x+9>5$$
$$\underset{-9}{}\underset{-9}{}$$

$$-4x>-4$$

$$\frac{-4x}{-4} < \frac{-4}{-4}$$

Division by a negative number reversed the inequality sign.

$x < 1$ which is the interval $(-\infty, 1)$.

5.012 $-2(5-3x)+1>18x+3(1-4x)$
$-10+6x+1>18x+3-12x$
$-9+6x>6x+3$
$\underset{-6x\quad\quad-6x}{-9+6x>6x+3}$

$-9>3$ which is *never* true. So the inequality has no solution.

5.013 $x+19-3(x+1)<2(x-8)$
$x+19-3x-3<2x-16$
$-2x+16<2x-16$
$\underset{-2x\quad\quad-2x}{-2x+16<2x-16}$
$-4x+16<-16$
$\underset{-16\quad\quad-16}{-4x+16<-16}$
$-4x<-32$
$\frac{-4x}{-4} > \frac{-32}{-4}$

Division by a negative number reversed the inequality sign.

$x > 8$ which is the interval $(8, \infty)$.

5.014 $x-2\big(x-2(x-2)\big)<3(x-2)$
$x-2(x-2x+4)<3x-6$
$x-2(-x+4)<3x-6$
$x+2x-8<3x-6$
$3x-8<3x-6$
$\underset{-3x\quad\quad-3x}{3x-8<3x-6}$

$-8<-6$ which is true regardless of the value of x. So x can take on absolutely any real number value. The solution is $\mathbb{R}$, or $(-\infty, \infty)$. The graph would be the entire number line as shown below.

5.020 $-2 \le 1-3x < 2$
$\underset{-1\quad-1\quad\quad-1}{\phantom{-2\le 1-3x<2}}$

$-3 \le -3x < 1$
$\frac{-3}{-3} \ge \frac{-3x}{-3} > \frac{1}{-3}$
$1 \ge x > -\frac{1}{3}$

The inequality shows values in decreasing order. They need to be in *increasing* order for us to correctly graph them on the number line. So, reverse the values and inequality signs, and rewrite.

$-\frac{1}{3} < x \le 1$ which is $\left(-\frac{1}{3}, 1\right]$.

The graph is shown below.

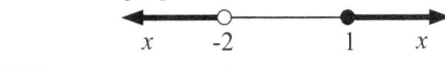

5.021 $x+3 \ge 4$ OR $8-x>10$
$\underset{-3\quad-3}{}$ $\underset{-8\quad\quad-8}{}$
$x \ge 1$ OR $-x>2$
$x \ge 1$ OR $\frac{-x}{-1} < \frac{2}{-1}$
$x \ge 1$ OR $x<-2$
$x<-2$ OR $x \ge 1$

The graph is shown below.

5.022 $0 < 2(x+11) \le 50$
$0 < 2x+22 \le 50$
$\underset{-22\quad\quad-22\quad-22}{0 < 2x+22 \le 50}$
$-22 < 2x \le 28$
$\frac{-22}{2} < \frac{2x}{2} \le \frac{28}{2}$
$-11 < x \le 14$

The graph is shown below.

5.023 $\frac{2+3x}{4} \cdot 4 \le 5 \cdot 4$ OR $5x+5 \ge 50$
$\underset{-5\quad-5}{}$
$2+3x \le 20$ OR $5x \ge 45$
$\underset{-2\quad\quad-2}{2+3x \le 20}$ OR $\frac{5x}{5} \ge \frac{45}{5}$
$3x \le 18$ OR $x \ge 9$
$\frac{3x}{3} \le \frac{18}{3}$ OR $x \ge 9$
$x \le 6$ OR $x \ge 9$

The graph is shown below.

5.024
$$2 \cdot 6 < \frac{3x-12}{6} \cdot 6 < 3 \cdot 6$$
$$12 < 3x - 12 < 18$$
$$\underset{+12 \quad +12 \quad +12}{12 < 3x - 12 < 18}$$
$$24 < 3x < 30$$
$$\frac{24}{3} < \frac{3x}{3} < \frac{30}{3}$$
$$8 < x < 10$$

The graph is shown below.

The interval is $(8,10)$.

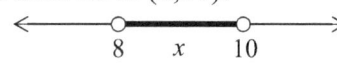

5.029
$$\underset{-1 \quad -1}{|5x-13|+1<18}$$
$$|5x-13|<17$$

The modulus is positive. The inequality is valid. Write the two separate inequalities:

$5x-13<17$	$-(5x-13)<17$
$\underset{+13 \quad +13}{5x-13<17}$	$-5x+13<17$
$5x<30$	$\underset{-13 \quad -13}{-5x+13<17}$
$\frac{5x}{5} < \frac{30}{5}$	$-5x<4$
$x<6$	$\frac{-5x}{-5} > \frac{4}{-5}$
	$x > -\frac{4}{5}$

The graph is shown below.

The interval is $\left(-\frac{4}{5}, 6\right)$.

5.030
$$\underset{-8 \quad -8}{|4-3x|+8<6}$$
$$|4-3x|<-2$$

The modulus is negative. The inequality is invalid, and has no solution.

5.031
$$\underset{-7 \quad\quad -7}{7+|3x-1|\geq 21}$$
$$|3x-1|\geq 14$$

The modulus is positive. The inequality is valid. Write the two separate inequalities:

$3x-1\geq 14$	$-(3x-1)\geq 14$
$\underset{+1 \quad +1}{3x-1\geq 14}$	$-3x+1\geq 14$

$3x\geq 15$	$\underset{-1 \quad -1}{-3x+1\geq 14}$
$\frac{3x}{3} \geq \frac{15}{3}$	$-3x\geq 13$
$x\geq 5$	$\frac{-3x}{-3} \leq \frac{13}{-3}$
	$x \leq -\frac{13}{3}$

The graph is shown below.

The interval is $\left(-\infty, -\frac{13}{3}\right] \cup [5,\infty)$.

5.032 $|x+6|\leq 9$

The modulus is positive. The inequality is valid. Write the two separate inequalities:

$x+6\leq 9$	$-(x+6)\leq 9$
$\underset{-6 \quad -6}{x+6\leq 9}$	$-x-6\leq 9$
$x\leq 3$	$\underset{+6 \quad +6}{-x-6\leq 9}$
	$-x\leq 15$
	$\frac{-x}{-1} \geq \frac{15}{-1}$
	$x\geq -15$

The graph is shown below.

The interval is $[-15,3]$.

5.037 $y \geq \frac{1}{4}x$

In the slope of the line, the run is 4. So go 4 paces to the left and right of 0 on the x-axis. You'll find the coordinates -4 and $+4$. Calculate the y-coordinates at $x=-4$, $+4$, and 0 thinking of the inequality as an *equation*. Plot those points and draw a *solid* line through them. Shade the region *above* it. The table is shown below, and the graph is on the next page.

x	$y=\frac{1}{4}x$
-4	$y=\frac{1}{4}\cdot(-4)=\frac{-4}{4}=-1$
0	$y=\frac{1}{4}\cdot(0)=0$
4	$y=\frac{1}{4}\cdot(4)=\frac{4}{4}=1$

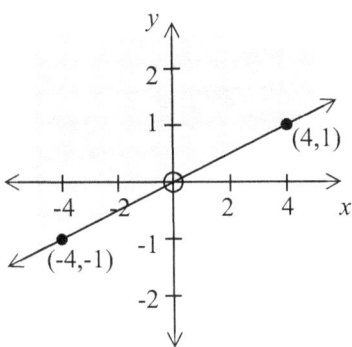

5.038 First solve the inequality for y.

$$4x - 2y < 2$$
$$\underset{-4x}{} \qquad \underset{-4x}{}$$
$$-2y < 2 - 4x$$
$$-2y < -4x + 2$$
$$\frac{-2y}{-2} > \frac{-4x+2}{-2}$$
$$y > \frac{2}{1}x - 1$$

In the slope of the line, the run is 1. So go 1 step to the left and right of 0 on the x-axis. You'll find the coordinates -1 and $+1$. Calculate the y-coordinates at $x = -1$, $+1$, and 0 thinking of the given inequality as an *equation*. Plot those points and draw a *dashed* line through them. Shade the region *above* it. Table and graph follow.

x	$y = 2x - 1$
-1	$y = 2 \cdot (-1) - 1 = -2 - 1 = -3$
0	$y = 2 \cdot (0) - 1 = 0 - 1 = -1$
1	$y = 2 \cdot (1) - 1 = 2 - 1 = 1$

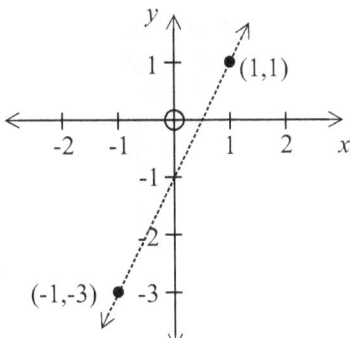

5.039 $\quad y \le -\dfrac{3}{4}x + 2$

In the slope of the line, the run is 4. So go 4 paces to the left and right of 0 on the x-axis. You'll find the coordinates -4 and $+4$. Calculate the y-coordinates at $x = -4$, $+4$, and 0 thinking

of the inequality as an *equation*. Plot those points and draw a *solid* line through them. Shade the region *below* it. Table and graph follow.

x	$y = -\dfrac{3}{4}x + 2$
-4	$y = -\dfrac{3}{4} \cdot (-4) + 2 = \dfrac{12}{4} + 2 = 3 + 2 = 5$
0	$y = -\dfrac{3}{4} \cdot (0) + 2 = 0 + 2 = 2$
4	$y = -\dfrac{3}{4} \cdot 4 + 2 = \dfrac{-12}{4} + 2 = -3 + 2 = -1$

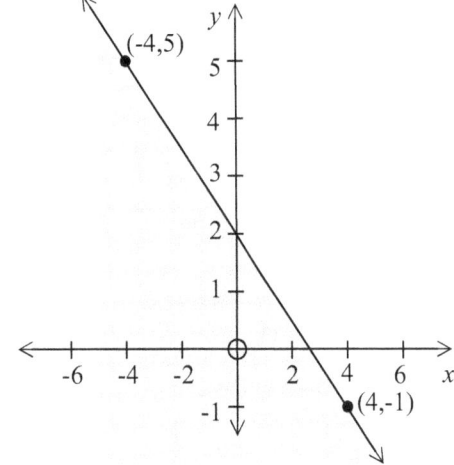

5.040 First solve the inequality for y.

$$y - x < 3$$
$$\underset{+x}{} \quad \underset{+x}{}$$
$$y < 3 + x$$
$$y < x + 3$$
$$y < \frac{1}{1}x + 3$$

In the slope of the line, the run is 1. So go 1 step to the left and right of 0 on the x-axis. You'll find the coordinates -1 and $+1$. Calculate the y-coordinates at $x = -1$, $+1$, and 0 thinking of the given inequality as an *equation*. Plot those points and draw a *dashed* line through them. Shade the region *below* it. The table is shown below, and the graph is on the next page.

x	$y = x + 3$
-1	$y = -1 + 3 = 2$
0	$y = 0 + 3 = 3$
1	$y = 1 + 3 = 4$

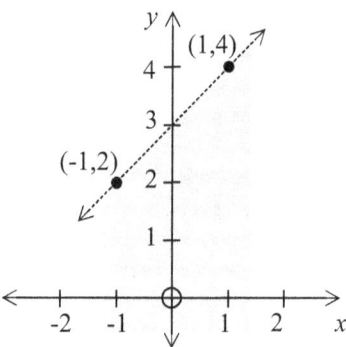

5.041 Isolate x, step by step.

$14 - 3x \geq 20$
$ -14 -14$
$-3x \geq 6$
$\dfrac{-3x}{-3} \leq \dfrac{6}{-3}$

Division by a negative number reversed the inequality sign.

$x \leq -2$ which is the interval $(-\infty, -2]$.

The graph is shown below.

5.042 Distribute and simplify.

$x + 5(x - 1) \leq 3(2x - 1)$
$x + 5x - 5 \leq 6x - 3$
$6x - 5 \leq 6x - 3$
$6x - 5 \leq 6x - 3$
$ -6x -6x$
$-5 \leq -3$

This inequality is *true* regardless of the value of x. Meaning, x can be any real number, the solution being $\mathbb{R}$, or the interval $(-\infty, \infty)$. The graph would be the entire number line as shown below.

5.043 Distribute and simplify.

$2x + 8 > 2(x + 4)$
$2x + 8 > 2x + 8$
$ -8 -8$
$2x > 2x$
$2x > 2x$
$-2x -2x$
$0 > 0$

The inequality is invalid, because 0 is not greater than 0. No solution.

5.044 Isolate x, step by step.

$-6 \leq \dfrac{x + 5}{3} < 6$

$-6 \cdot 3 \leq \dfrac{x + 5}{3} \cdot 3 < 6 \cdot 3$

$-18 \leq x + 5 < 18$

$-18 \leq x + 5 < 18$
$ -5 -5 \phantom{<1} -5$

$-23 \leq x < 13$ which is the interval $[-23, 13)$.

The graph is shown below.

5.045 Isolate x in both inequalities, step by step.

$\dfrac{x - 4}{9} \leq 1$ OR $2x - 14 \geq 26$

$\dfrac{x - 4}{9} \cdot 9 \leq 1 \cdot 9$ OR $2x - 14 \geq 26$
$\phantom{\dfrac{x-4}{9} \cdot 9 \leq 1 \cdot 9 \text{ OR } 2x-14}{+14} {+14}$

$x - 4 \leq 9$ OR $2x \geq 40$

$x - 4 \leq 9$ OR $\dfrac{2x}{2} \geq \dfrac{40}{2}$
${+4} {+4}$

$x \leq 13$ OR $x \geq 20$

The interval is $(-\infty, 13] \cup [20, \infty)$.

The graph is shown below.

5.046 First isolate the modulus.

$5|4x - 6| + 13 < 8$
$5|4x - 6| + 13 < 8$
${-13} \phantom{<} {-13}$
$5|4x - 6| < -5$
$\dfrac{5|4x - 6|}{5} < \dfrac{-5}{5}$
$|4x - 6| < -1$

The modulus is negative. The inequality is invalid, and has no solution.

5.047 $|9 - 3x| < 6$

Write the two separate inequalities:

$9 - 3x < 6$	$-(9 - 3x) < 6$
$9 - 3x < 6$	$-9 + 3x < 6$
$ {-9} \phantom{-3x<} {-9}$	
$-3x < -3$	$-9 + 3x < 6$
	$ {+9} \phantom{+3x<} {+9}$
$\dfrac{-3x}{-3} > \dfrac{-3}{-3}$	$3x < 15$
$x > 1$	$\dfrac{3x}{3} < \dfrac{15}{3}$
	$x < 5$

The graph is shown below.
The interval is (1,5).

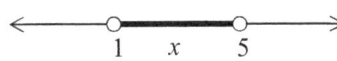

5.048 First isolate the modulus.

$$2|x+7|-1\leq9$$

$$2|x+7|\underset{+1\quad+1}{-1\leq9}$$

$$2|x+7|\leq10$$

$$\frac{2|x+7|}{2}\leq\frac{10}{2}$$

$$|x+7|\leq5$$

The modulus is positive. The inequality is valid. Write the two separate inequalities:

$$
\begin{array}{c|c}
x+7\leq5 & -(x+7)\leq5 \\
\underset{-7\quad-7}{x+7\leq5} & -x-7\leq5 \\
x\leq-2 & \underset{+7\quad+7}{-x-7\leq5} \\
 & -x\leq12 \\
 & \dfrac{-x}{-1}\geq\dfrac{12}{-1} \\
 & x\geq-12
\end{array}
$$

The graph is shown below.
The interval is [−12,−2].

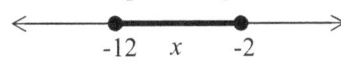

5.049 $y>-\dfrac{1}{2}x+2$

In the slope of the line, the run is 2. So go 2 paces to the left and right of 0 on the *x*-axis. You'll find the coordinates −2 and +2. Calculate the *y*-coordinates at $x=-2$, $+2$, and 0 thinking of the inequality as an *equation*. Plot those points and draw a *dashed* line through them. Shade the region *above* it. Table and graph follow.

x	$y=-\dfrac{1}{2}x+2$
−2	$y=-\dfrac{1}{2}\cdot(-2)+2=\dfrac{2}{2}+2=1+2=3$
0	$y=-\dfrac{1}{2}\cdot0+2=0+2=2$
2	$y=-\dfrac{1}{2}\cdot2+2=\dfrac{-2}{2}+2=-1+2=1$

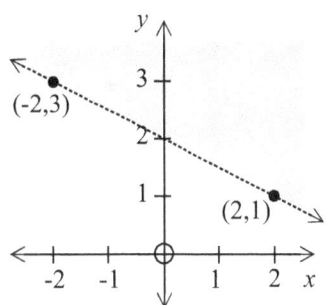

5.050 First solve the inequality for *y*.

$$3x+y\leq1$$

$$\underset{-3x\qquad-3x}{3x+y\leq1}$$

$$y\leq1-3x$$

$$y\leq-3x+1$$

$$y\leq-\frac{3}{1}x+1$$

In the slope of the line, the run is 1. So go 1 step to the left and right of 0 on the *x*-axis. You'll find the coordinates −1 and +1. Find the *y*-coordinates at $x=-1$, $+1$, and 0 thinking of the given inequality as an *equation*. Plot those points and draw a *solid* line through them. Shade the region *below* it. Table and graph follow.

x	$y=-3x+1$
−1	$y=-3\cdot(-1)+1=3+1=4$
0	$y=-3\cdot(0)+1=0+1=1$
1	$y=-3(1)+1=-3+1=-2$

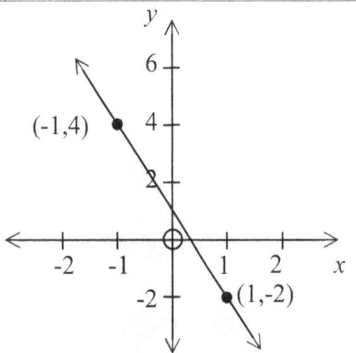

CHAPTER 6: SEQUENCES AND SERIES

6.005 22, 19, 16, 13, …

$$d=a_2-a_1=19-22=-3$$

$$a_n=a_1+d(n-1)$$

$$a_{15}=22+(-3)(15-1)$$

$$a_{15}=22-3(14)$$

$$a_{15}=-20$$

$$a_{50}=22+(-3)(50-1)$$

$a_{50}=22-3(49)$

$a_{50}=-125$

6.006 $-9, -5, -1, 3, \ldots$
$d=a_2-a_1=-5-(-9)=-5+9=4$
$a_n=a_1+d(n-1)$
$a_8=-9+4(8-1)$
$a_8=-9+4(7)$
$a_8=19$
$a_{23}=-9+4(23-1)$
$a_{23}=-9+4(22)$
$a_{23}=79$

6.007 $1.4, 2.7, 4, 5.3, \ldots$
$d=a_2-a_1=2.7-1.4=1.3$
$a_n=a_1+d(n-1)$
$a_9=1.4+1.3(9-1)$
$a_9=1.4+1.3(8)$
$a_9=11.8$
$a_{13}=1.4+1.3(13-1)$
$a_{13}=1.4+1.3(12)$
$a_{13}=17$

6.008 $64, 57, 50, 43, \ldots$
$d=a_2-a_1=57-64=-7$
$a_n=a_1+d(n-1)$
$a_{25}=64+(-7)(25-1)$
$a_{25}=64-7(24)$
$a_{25}=-104$
$a_{33}=64+(-7)(33-1)$
$a_{33}=64-7(32)$
$a_{33}=-160$

6.012 In example 6.001, the common difference is 3. To obtain the nth term (a_n) add 3 to the $(n-1)^{th}$ term (a_{n-1}). Recursive formula: $a_n=a_{n-1}+3$

In example 6.002, the common difference is 2. To obtain the nth term (a_n) add 2 to the $(n-1)^{th}$ term (a_{n-1}). Recursive formula: $a_n=a_{n-1}+2$

In example 6.003, the common difference is -4. To obtain the nth term (a_n) add -4 to the $(n-1)^{th}$ term (a_{n-1}). Recursive formula: $a_n=a_{n-1}+(-4)$ or simply $a_n=a_{n-1}-4$

In example 6.004, common difference is -0.25. To obtain the nth term (a_n) add -0.25 to the $(n-1)^{th}$ term (a_{n-1}). Recursive formula:
$a_n=a_{n-1}+(-0.25)$ or simply $a_n=a_{n-1}-0.25$

6.013 $68, 72, 76, 80, \ldots$ The common difference is: $d=a_2-a_1=72-68=4$ To obtain the nth term (a_n) add 4 to the $(n-1)^{th}$ term (a_{n-1}). Recursive formula: $a_n=a_{n-1}+4$

6.014 $1.2, 0.7, 0.2, -0.3, \ldots$ The common difference is: $d=a_2-a_1=0.7-1.2=-0.5$ To obtain the nth term (a_n) add -0.5 to the $(n-1)^{th}$ term (a_{n-1}). Recursive formula: $a_n=a_{n-1}-0.5$

6.019 $2, 6, 10, 14, \ldots$ Go in the order d, a_n, S_n.
$d=a_2-a_1=6-2=4$
$a_n=a_1+d(n-1)$
$a_{11}=2+4(11-1)$
$a_{11}=2+4(10)$
$a_{11}=42$
$S_n=\left(\dfrac{a_1+a_n}{2}\right)\cdot n$
$S_{11}=\left(\dfrac{a_1+a_{11}}{2}\right)\cdot 11$
$S_{11}=\left(\dfrac{2+42}{2}\right)\cdot 11$
$S_{11}=242$

6.020 $4.6 + 4.1 + 3.6 + 3.1 + \ldots$
Go in the order d, a_n, S_n.
$d=a_2-a_1=4.1-4.6=-0.5$
$a_n=a_1+d(n-1)$
$a_{20}=4.6+(-0.5)(20-1)$
$a_{20}=4.6-0.5(19)$
$a_{20}=-4.9$
$S_n=\left(\dfrac{a_1+a_n}{2}\right)\cdot n$
$S_{20}=\left(\dfrac{a_1+a_{20}}{2}\right)\cdot 20$
$S_{20}=\left(\dfrac{4.6+(-4.9)}{2}\right)\cdot 20$
$S_{20}=\left(\dfrac{-0.3}{2}\right)\cdot 20=-3$

6.021 $-7.3,\ -7.1,\ -6.9,\ -6.7,\ \ldots$

Go in the order d, a_n, S_n.

$$d = a_2 - a_1 = -7.1 - (-7.3) = -7.1 + 7.3 = 0.2$$

$$a_n = a_1 + d(n-1)$$

$$a_{17} = -7.3 + 0.2(17-1)$$

$$a_{17} = -7.3 + 0.2(16)$$

$$a_{17} = -4.1$$

$$S_n = \left(\frac{a_1 + a_n}{2}\right) \cdot n$$

$$S_{17} = \left(\frac{a_1 + a_{17}}{2}\right) \cdot 17$$

$$S_{17} = \left(\frac{-7.3 + (-4.1)}{2}\right) \cdot 17$$

$$S_{17} = \left(\frac{-11.4}{2}\right) \cdot 17 = -96.9$$

6.022 $(-32) + (-34) + (-36) + (-38) + \ldots$

Go in the order d, a_n, S_n.

$$d = a_2 - a_1 = -34 - (-32) = -34 + 32 = -2$$

$$a_n = a_1 + d(n-1)$$

$$a_8 = -32 + (-2)(8-1)$$

$$a_8 = -32 - 2(7)$$

$$a_8 = -46$$

$$S_n = \left(\frac{a_1 + a_n}{2}\right) \cdot n$$

$$S_8 = \left(\frac{a_1 + a_8}{2}\right) \cdot 8$$

$$S_8 = \left(\frac{-32 + (-46)}{2}\right) \cdot 8$$

$$S_8 = \left(\frac{-78}{2}\right) \cdot 8 = -312$$

6.027 $\dfrac{1}{2}, -\dfrac{1}{3}, \dfrac{2}{9}, -\dfrac{4}{27}, \ldots$

$$r = a_2 \div a_1 = -\frac{1}{3} \div \frac{1}{2} = -\frac{1}{3} \times \frac{2}{1} = -\frac{2}{3}$$

Recursive formula: $a_n = a_{n-1} \times \left(-\dfrac{2}{3}\right)$

$$a_n = a_1 \cdot r^{n-1}$$

$$a_6 = \frac{1}{2} \cdot \left(-\frac{2}{3}\right)^{6-1} \qquad a_9 = \frac{1}{2} \cdot \left(-\frac{2}{3}\right)^{9-1}$$

$$a_6 = \frac{1}{2} \cdot \left(-\frac{2}{3}\right)^{5} = -\frac{16}{243} \qquad a_9 = \frac{1}{2} \cdot \left(-\frac{2}{3}\right)^{8} = \frac{128}{6561}$$

6.028 $\dfrac{1}{4}, \dfrac{1}{5}, \dfrac{4}{25}, \dfrac{16}{125}, \ldots$

$$r = a_2 \div a_1 = \frac{1}{5} \div \frac{1}{4} = \frac{1}{5} \times \frac{4}{1} = \frac{4}{5}$$

Recursive formula: $a_n = a_{n-1} \times \dfrac{4}{5}$

The next three terms would be:

$$a_5 = a_4 \times \frac{4}{5} = \frac{16}{125} \times \frac{4}{5} = \frac{64}{625}$$

$$a_6 = a_5 \times \frac{4}{5} = \frac{64}{625} \times \frac{4}{5} = \frac{256}{3125}$$

$$a_7 = a_6 \times \frac{4}{5} = \frac{256}{3125} \times \frac{4}{5} = \frac{1024}{15625}$$

6.029 $1,\ 1.1,\ 1.21,\ 1.331,\ \ldots$

$$r = \frac{a_2}{a_1} = \frac{1.1}{1} = 1.1$$

$$a_n = a_1 \cdot r^{n-1}$$

$$a_6 = 1 \cdot (1.1)^{6-1}$$

$$a_6 = 1 \cdot (1.1)^{5} = 1.61051$$

$$a_8 = 1 \cdot (1.1)^{8-1}$$

$$a_8 = 1 \cdot (1.1)^{7} = 1.9487171$$

6.030 $-100,\ 20,\ -4,\ 0.8,\ \ldots$

$$r = \frac{a_2}{a_1} = \frac{20}{-100} = \frac{1}{-5} = -0.2$$

$$a_n = a_1 \cdot r^{n-1}$$

$$a_7 = -100 \cdot (-0.2)^{7-1}$$

$$a_7 = -100 \cdot (-0.2)^{6}$$

$$a_7 = -0.0064$$

6.035 $(-7) + 7 + (-7) + 7 + \ldots$

It's given that $a_1 = -7$, $n = 200$ (for S_{200}).

$$r = \frac{a_2}{a_1} = \frac{7}{-7} = -1$$

$$S_n = a_1 \cdot \left(\frac{1 - r^n}{1 - r}\right)$$

$$S_{200} = -7 \cdot \left(\frac{1 - (-1)^{200}}{1 - (-1)}\right)$$

$$S_{200} = -7 \cdot \left(\frac{1 - 1}{1 + 1}\right) = -7 \cdot \left(\frac{0}{2}\right) = 0$$

6.036 $\dfrac{1}{125}, \dfrac{1}{25}, \dfrac{1}{5}, 1, \ldots$

It's given that $a_1 = \dfrac{1}{125}$, $n = 8$.

$$r = a_2 \div a_1 = \frac{1}{25} \div \frac{1}{125} = \frac{1}{25} \times \frac{125}{1} = \frac{125}{25} = 5$$

$$S_n = a_1 \cdot \left(\frac{1 - r^n}{1 - r} \right)$$

$$S_8 = \frac{1}{125} \cdot \left(\frac{1 - 5^8}{1 - 5} \right)$$

$$S_8 = \frac{1}{125} \cdot \left(\frac{-390624}{-4} \right) = \frac{97656}{125} = 781.248$$

6.037 $3 + 3^2 + 3^3 + 3^4 + \ldots$

It's given that $a_1 = 3$, $n = 6$.

$$r = \frac{a_2}{a_1} = \frac{3^2}{3} = 3$$

$$S_n = a_1 \cdot \left(\frac{1 - r^n}{1 - r} \right)$$

$$S_6 = 3 \cdot \left(\frac{1 - 3^6}{1 - 3} \right)$$

$$S_6 = 3 \cdot \left(\frac{-728}{-2} \right) = 1092$$

6.038 $3072, -1536, 768, -384, \ldots$

It's given that $a_1 = 3072$, $n = 13$.

$$r = \frac{a_2}{a_1} = \frac{-1536}{3072} = \frac{-1}{2} = -0.5$$

$$S_n = a_1 \cdot \left(\frac{1 - r^n}{1 - r} \right)$$

$$S_{13} = 3072 \cdot \left(\frac{1 - (-0.5)^{13}}{1 - (-0.5)} \right) = 2048.25$$

6.042 For a series made with example 6.038, the common ratio is:

$$r = \frac{a_2}{a_1} = \frac{-1536}{3072} = \frac{-1}{2} = -0.5$$

This r value is within the interval $(-1, 1)$. The series is therefore convergent.

$$S_n = a_1 \cdot \left(\frac{1 - r^n}{1 - r} \right)$$

$$S_\infty = 3072 \cdot \left(\frac{1 - (-0.5)^\infty}{1 - (-0.5)} \right)$$

$$S_\infty = 3072 \cdot \left(\frac{1 - 0}{1 + 0.5} \right) = 3072 \cdot \left(\frac{1}{1.5} \right) = 2048$$

6.043 $\dfrac{1}{5} + \dfrac{1}{5^2} + \dfrac{1}{5^3} + \dfrac{1}{5^4} + \ldots$ The common ratio is

$r = a_2 \div a_1 = \dfrac{1}{5^2} \div \dfrac{1}{5} = \dfrac{1}{5^2} \times \dfrac{5}{1} = \dfrac{1}{5} = 0.2$ which is in

the interval $(-1, 1)$. The series is therefore convergent.

$$S_n = a_1 \cdot \left(\frac{1 - r^n}{1 - r} \right)$$

$$S_\infty = \frac{1}{5} \cdot \left(\frac{1 - \left(\dfrac{1}{5} \right)^\infty}{1 - \dfrac{1}{5}} \right)$$

$$S_\infty = \frac{1}{5} \cdot \left(\frac{1 - 0}{\dfrac{4}{5}} \right) = \frac{1}{5} \cdot \left(\frac{1}{\dfrac{4}{5}} \right) = \frac{1}{5} \cdot 1 \cdot \frac{5}{4} = \frac{1}{4} = 0.25$$

6.044 $2.5, 2.9, 3.3, 3.7, \ldots$ $\quad a_1 = 2.5$, $n = 25$

Common difference $d = a_2 - a_1 = 2.9 - 2.5 = 0.4$

$$a_n = a_1 + d(n - 1)$$
$$a_{25} = 2.5 + 0.4(25 - 1)$$
$$a_{25} = 2.5 + 0.4(24) = 12.1$$

Recursive formula: $\quad a_n = a_{n-1} + 0.4$

6.045 $99, 84, 79, 64, \ldots$ $\quad a_1 = 99$, $n = 20$

Common difference $d = a_2 - a_1 = 84 - 99 = -15$

$$a_n = a_1 + d(n - 1)$$
$$a_{20} = 99 + (-15)(20 - 1)$$
$$a_{20} = 99 - 15(19) = -186$$

Recursive formula: $\quad a_n = a_{n-1} - 15$

6.046 $7 + 10 + 13 + 16 + \ldots$ $\quad a_1 = 7$, $n = 50$

Common difference $d = a_2 - a_1 = 10 - 7 = 3$

$$a_n = a_1 + d(n - 1)$$
$$a_{50} = 7 + 3(50 - 1)$$
$$a_{50} = 7 + 3(49) = 154$$

$$S_n = \left(\frac{a_1 + a_n}{2} \right) \cdot n$$

$$S_{50} = \left(\frac{a_1 + a_{50}}{2} \right) \cdot 50$$

$$S_{50} = \left(\frac{7 + 154}{2} \right) \cdot 50 = \left(\frac{161}{2} \right) \cdot 50 = 4025$$

6.047 $-1, -7, -13, -19, \dots$ $a_1 = -1, \; n = 15$

Common difference is

$d = a_2 - a_1 = -7 - (-1) = -7 + 1 = -6$

$a_n = a_1 + d(n-1)$

$a_{15} = -1 + (-6)(15-1)$

$a_{15} = -1 - 6(14) = -85$

$S_n = \left(\dfrac{a_1 + a_n}{2}\right) \cdot n$

$S_{15} = \left(\dfrac{a_1 + a_{15}}{2}\right) \cdot 15$

$S_{15} = \left(\dfrac{-1 + (-85)}{2}\right) \cdot 15$

$S_{15} = \left(\dfrac{-1 - 85}{2}\right) \cdot 15 = \left(\dfrac{-86}{2}\right) \cdot 15 = -645$

6.048 $6, 12, 24, 48, \dots$ $a_1 = 6, \; n = 9$

Common ratio $r = \dfrac{a_2}{a_1} = \dfrac{12}{6} = 2$

$a_n = a_1 \cdot r^{n-1}$

$a_9 = 6 \cdot 2^{9-1} = 6 \cdot 2^8 = 6 \cdot 256 = 1536$

Recursive formula: $a_n = a_{n-1} \times 2$

6.049 $63, \; -21, \; 7, \; -\dfrac{7}{3}, \; \dots$ $a_1 = 63, \, n = 7$

Common ratio $r = \dfrac{a_2}{a_1} = \dfrac{-21}{63} = \dfrac{-1}{3} = -\dfrac{1}{3}$

$a_n = a_1 \cdot r^{n-1}$

$a_7 = 63 \cdot \left(-\dfrac{1}{3}\right)^{7-1} = 63 \cdot \left(-\dfrac{1}{3}\right)^6 = \dfrac{63}{729} = \dfrac{7}{81}$

Recursive formula: $a_n = a_{n-1} \times \left(-\dfrac{1}{3}\right)$

6.050 $30 + (-3) + 0.3 + (-0.03) + \dots$

$a_1 = 30, \; n = 8, \; r = \dfrac{a_2}{a_1} = \dfrac{-3}{30} = \dfrac{-1}{10} = -0.1$

$S_n = a_1 \cdot \left(\dfrac{1 - r^n}{1 - r}\right)$

$S_8 = 30 \cdot \left(\dfrac{1 - (-0.1)^8}{1 - (-0.1)}\right) = 27.272727$

6.051 $1792 + (-448) + 112 + (-28) + \dots$

$a_1 = 1792, \; n = 5, \; r = \dfrac{a_2}{a_1} = \dfrac{-448}{1792} = \dfrac{-1}{4} = -0.25$

$S_n = a_1 \cdot \left(\dfrac{1 - r^n}{1 - r}\right)$

$S_5 = 1792 \cdot \left(\dfrac{1 - (-0.25)^5}{1 - (-0.25)}\right) = 1435$

6.052 $\dfrac{1}{3} - \dfrac{1}{3^2} + \dfrac{1}{3^3} - \dfrac{1}{3^4} + \dots$ $a_1 = \dfrac{1}{3}, \; n = \infty$

$r = a_2 \div a_1 = \dfrac{-1}{3^2} \div \dfrac{1}{3} = \dfrac{-1}{3^2} \times \dfrac{3}{1} = \dfrac{-1}{3} = -\dfrac{1}{3}$

Common ratio (r) is within the interval $(-1, 1)$. The series is therefore convergent. Its sum is:

$S_n = a_1 \cdot \left(\dfrac{1 - r^n}{1 - r}\right)$

$S_\infty = \dfrac{1}{3} \cdot \left(\dfrac{1 - \left(-\dfrac{1}{3}\right)^\infty}{1 - \left(-\dfrac{1}{3}\right)}\right)$

$S_\infty = \dfrac{1}{3} \cdot \left(\dfrac{1 - 0}{1 + \dfrac{1}{3}}\right) = \dfrac{1}{3} \cdot \left(\dfrac{1}{\dfrac{4}{3}}\right) = \dfrac{1}{3} \cdot \dfrac{3}{4} = \dfrac{1}{4}$

6.053 $\dfrac{1}{6} + \dfrac{1}{6^2} + \dfrac{1}{6^3} + \dfrac{1}{6^4} + \dots$ $a_1 = \dfrac{1}{6}, \; n = \infty$

$r = a_2 \div a_1 = \dfrac{1}{6^2} \div \dfrac{1}{6} = \dfrac{1}{6^2} \times \dfrac{6}{1} = \dfrac{1}{6}$

Common ratio (r) is within the interval $(-1, 1)$. So the series is convergent. It converges on:

$S_n = a_1 \cdot \left(\dfrac{1 - r^n}{1 - r}\right)$

$S_\infty = \dfrac{1}{6} \cdot \left(\dfrac{1 - \left(\dfrac{1}{6}\right)^\infty}{1 - \dfrac{1}{6}}\right)$

$S_\infty = \dfrac{1}{6} \cdot \left(\dfrac{1 - 0}{\dfrac{5}{6}}\right) = \dfrac{1}{6} \cdot \left(\dfrac{1}{\dfrac{5}{6}}\right) = \dfrac{1}{6} \cdot \dfrac{6}{5} = \dfrac{1}{5}$

CHAPTER 7: COMPLEX NUMBERS

7.007 $\overbrace{(6 + 2i)} - \overbrace{(18 - i)}$

$\underbrace{6 - 18} \qquad \underbrace{+2i - (-i)}$

$-12 \quad \overbrace{+2i + i}$

$-12 + 3i$

7.008
$(-1+4i)+(17-3i)$

$\underbrace{-1+17}\quad\underbrace{+4i+(-3i)}$

$16\quad\underbrace{+4i-3i}$

$16+i$

7.009
$(2i-14)+(20-2i)$

$\underbrace{2i+(-2i)}\quad\underbrace{-14+20}$

$\underbrace{2i-2i}\quad+6$

$0i+6$

6

7.010
$(7+6i)-(-5i+7)$

$\underbrace{7-7}\quad\underbrace{+6i-(-5i)}$

$0\quad\underbrace{+6i+5i}$

$11i$

7.011
$(-i+11)+(-2i)$

$\underbrace{-i+(-2i)}\quad+11$

$\underbrace{-i-2i}\quad+11$

$-3i+11$

$11-3i$

7.012
$(-9+3i)-(10-7i)$

$\underbrace{-9-10}\quad\underbrace{+3i-(-7i)}$

$-19\quad\underbrace{+3i+7i}$

$-19+10i$

7.017
$(1-8i)(5+3i)$

$1\cdot5+1\cdot(3i)-(8i)\cdot5-(8i)\cdot(3i)$

$5+3i-40i-24i^2$

Replace i^2 with -1.

$5-37i-24(-1)$

$5-37i+24$

$29-37i$

7.018
$(-7i+6)(2-5i)$

$-(7i)\cdot2-(7i)\cdot(-5i)+6\cdot2+6\cdot(-5i)$

$-14i+35i^2+12-30i$

Replace i^2 with -1.

$-44i+35(-1)+12$

$-44i-35+12$

$-44i-23$

$-23-44i$

7.019
$(4+2i)(-4+2i)$

$4\cdot(-4)+4\cdot(2i)+(2i)\cdot(-4)+(2i)\cdot(2i)$

$-16+8i-8i+4i^2$

Replace i^2 with -1.

$-16+4(-1)$

$-16-4$

-20

7.020
$(3+25i)(1-i)$

$3\cdot1+3\cdot(-i)+(25i)\cdot1+(25i)\cdot(-i)$

$3-3i+25i-25i^2$

Replace i^2 with -1.

$3+22i-25(-1)$

$3+22i+25$

$28+22i$

7.025
$\dfrac{1+4i}{5-7i}$

Complex conjugate of denominator is $5+7i$. Multiply the numerator and denominator by it.

$\dfrac{1+4i}{5-7i}\times\dfrac{5+7i}{5+7i}$

$\dfrac{(1+4i)\cdot(5+7i)}{(5-7i)\cdot(5+7i)}$

$\dfrac{5+7i+20i+28i^2}{25+35i-35i-49i^2}$

Replace i^2 with -1.

$\dfrac{5+27i+28(-1)}{25-49(-1)}$

$\dfrac{5+27i-28}{25+49}$

$\dfrac{-23+27i}{74}$

$\dfrac{-23}{74}+\dfrac{27}{74}i$

7.026
$\dfrac{9-36i}{-9i}$

Complex conjugate of denominator is $9i$. Multiply the numerator and denominator by it.

$\dfrac{9-36i}{-9i}\times\dfrac{9i}{9i}$

$\dfrac{(9-36i)\cdot(9i)}{(-9i)\cdot(9i)}$

$\dfrac{81i-324i^2}{-81i^2}$

Replace i^2 with -1.

$$\frac{81i-324(-1)}{-81(-1)}$$

$$\frac{81i+324}{81}$$

$$\frac{81}{81}i+\frac{324}{81}$$

$i+4$

$4+i$

7.027 $\dfrac{2i-6}{-3+i}$

Complex conjugate of denominator is $-3-i$. Multiply the numerator and denominator by it.

$$\frac{2i-6}{-3+i}\times\frac{-3-i}{-3-i}$$

$$\frac{(2i-6)\cdot(-3-i)}{(-3+i)\cdot(-3-i)}$$

$$\frac{-6i-2i^2+18+6i}{9+3i-3i-i^2}$$

Replace i^2 with -1.

$$\frac{-2(-1)+18}{9-(-1)}$$

$$\frac{2+18}{9+1}$$

$$\frac{20}{10}$$

2

7.028 $\dfrac{-7+4i}{10-i}$

Complex conjugate of denominator is $10+i$. Multiply the numerator and denominator by it.

$$\frac{-7+4i}{10-i}\times\frac{10+i}{10+i}$$

$$\frac{(-7+4i)\cdot(10+i)}{(10-i)\cdot(10+i)}$$

$$\frac{-70-7i+40i+4i^2}{100+10i-10i-i^2}$$

Replace i^2 with -1.

$$\frac{-70+33i+4(-1)}{100-(-1)}$$

$$\frac{-70+33i-4}{100+1}$$

$$\frac{-74+33i}{101}$$

$$\frac{-74}{101}+\frac{33}{101}i$$

7.033 $(2i^5)^3$

$2^3\cdot(i^5)^3$ Refer page 43, law 7.

$8\cdot i^{15}$ Refer page 43, law 3.

$8\cdot i^{12+3}$

$8\cdot i^{12}\cdot i^3$ Refer page 43, law 1.

$8\cdot 1\cdot i^3$ because $i^{12}=i^4=1$

$8\cdot i^2\cdot i$

$8\cdot(-1)\cdot i$

$-8i$

7.034 i^{22}

i^{20+2}

$i^{20}\cdot i^2$ Refer page 43, law 1.

$1\cdot(-1)$ because $i^{20}=i^4=1$

-1

7.035 $(i^3+i)(i^{10}-i^6)$

$i^{3+10}-i^{3+6}+i^{1+10}-i^{1+6}$

$i^{13}-i^9+i^{11}-i^7$

$i^{12+1}-i^{8+1}+i^{8+3}-i^{4+3}$

$i^{12}\cdot i-i^8\cdot i+i^8\cdot i^3-i^4\cdot i^3$

$1\cdot i-1\cdot i+1\cdot i^3-1\cdot i^3$

$i-i+i^3-i^3$

0

7.036 $25i^6\cdot 13i^7$

$25\cdot 13\cdot i^{6+7}$

$325i^{13}$

$325i^{12+1}$

$325\cdot i^{12}\cdot i$

$325\cdot 1\cdot i$

$325i$

7.037 $(16+11i)+(-4-4i)$

Regroup the like terms.

$16-4+11i-4i$

$12+7i$

7.038 $(5-4i)-(-4i+7)$

Regroup the like terms, keeping in mind the havoc the negative sign can play!

$5-7-4i+4i$

-2

7.039 $(6i+8)+(6-4i)$

Regroup the like terms.

$8+6+6i-4i$

$14+2i$

7.040 $(5i+4)-(9i+4)$

Regroup the like terms.

$4-4+5i-9i$

$-4i$

7.041 $(2+i)(-8+3i)$

Distribute 2 onto $(-8+3i)$. Then distribute i onto $(-8+3i)$.

$-16+6i-8i-3i^2$

$-16-2i-3i^2$

$-16-2i-3(-1)$ because $i^2=-1$

$-16-2i+3$

$-13-2i$

7.042 $(4+i)(i-4)$

Distribute 4 onto $(i-4)$. Then distribute i onto $(i-4)$.

$4i-16+i^2-4i$

$-16+i^2$

$-16-1$ because $i^2=-1$

-17

7.043 $\dfrac{3-5i}{7+3i}$

Complex conjugate of denominator is $7-3i$. Multiply the numerator and denominator by it.

$\dfrac{3-5i}{7+3i}\times\dfrac{7-3i}{7-3i}$

$\dfrac{(3-5i)\cdot(7-3i)}{(7+3i)\cdot(7-3i)}$

$\dfrac{21-9i-35i+15i^2}{49-21i+21i-9i^2}$

$\dfrac{21-44i+15i^2}{49-9i^2}$

$\dfrac{21-44i+15(-1)}{49-9(-1)}$ because $i^2=-1$

$\dfrac{21-44i-15}{49+9}$

$\dfrac{6-44i}{58}$

$\dfrac{6}{58}-\dfrac{44}{58}i$ which is $\dfrac{3}{29}-\dfrac{22}{29}i$

7.044 $\dfrac{18i+12}{3i}$

Complex conjugate of denominator is $-3i$. Multiply the numerator and denominator by it.

$\dfrac{18i+12}{3i}\times\dfrac{-3i}{-3i}$

$\dfrac{(18i+12)\cdot(-3i)}{(3i)\cdot(-3i)}$

$\dfrac{-54i^2-36i}{-9i^2}$

$\dfrac{-54(-1)-36i}{-9(-1)}$ because $i^2=-1$

$\dfrac{54-36i}{9}$

$\dfrac{54}{9}-\dfrac{36}{9}i$

$6-4i$

7.045 $i^{13}-(i^5)^7$

$i^{12+1}-i^{5\times7}$ Refer page 43, law 3.

$i^{12+1}-i^{35}$

$i^{12+1}-i^{32+3}$

$i^{12}\cdot i-i^{32}\cdot i^3$ Refer page 43, law 1.

$1\cdot i-1\cdot i^3$ because $i^{12}=i^{32}=i^4=1$

$i-i^3$

$i-i^2\cdot i$

$i+i$ because $i^2=-1$

$2i$

7.046 $-6i(5i-i^3)$

$-30i^2+6i^4$

$-30\cdot(-1)+6\cdot1$ since $i^2=-1$; $i^4=1$

$30+6$

36

1	2	3	4	5	6	7	8	9	10	11	12	13	14	15
2	**4**	6	8	10	12	14	16	18	20	22	24	26	28	30
3	6	**9**	12	15	18	21	24	27	30	33	36	39	42	45
4	8	12	**16**	20	24	28	32	36	40	44	48	52	56	60
5	10	15	20	**25**	30	35	40	45	50	55	60	65	70	75
6	12	18	24	30	**36**	42	48	54	60	66	72	78	84	90
7	14	21	28	35	42	**49**	56	63	70	77	84	91	98	105
8	16	24	32	40	48	56	**64**	72	80	88	96	104	112	120
9	18	27	36	45	54	63	72	**81**	90	99	108	117	126	135
10	20	30	40	50	60	70	80	90	**100**	110	120	130	140	150
11	22	33	44	55	66	77	88	99	110	**121**	132	143	154	165
12	24	36	48	60	72	84	96	108	120	132	**144**	156	168	180
13	26	39	52	65	78	91	104	117	130	143	156	**169**	182	195
14	28	42	56	70	84	98	112	126	140	154	168	182	**196**	210
15	30	45	60	75	90	105	120	135	150	165	180	195	210	**225**

Note:
1) It would greatly help you if you memorize the multiples of at least the first 10 integers, highlighted in gray. If you can memorize more than that, then that's even better!
2) The numbers along the diagonal of the table are integer squares, **written in bold.**

www.ingramcontent.com/pod-product-compliance
Lightning Source LLC
Chambersburg PA
CBHW080010210526
45170CB00015B/1967